**Optical Refrigeration**

*Edited by*
*Richard Epstein and*
*Mansoor Sheik-Bahae*

## Related Titles

Capper, P., Mauk, M. (eds.)

**Liquid Phase Epitaxy of Electronic, Optical and Optoelectronic Materials**

2007
ISBN 978-0-470-85290-3

Incropera, F. P., DeWitt, D. P., Bergman, T. L., Lavine, A. S.

**Fundamentals of Heat and Mass Transfer**

2006
ISBN 978-0-471-45728-2

Csele, M.

**Fundamentals of Light Sources and Lasers**

2004
ISBN 978-0-471-47660-3

Dinçer, I.

**Refrigeration Systems and Applications**

2003
ISBN: 978-0-471-62351-9

Weidemüller, M., Zimmermann, C. (eds.)

**Cold Atoms and Molecules**
Concepts, Experiments and Applications to Fundamental Physics

2009
ISBN 978-3-527-40750-7

Weidemüller, M., Zimmermann, C. (eds.)

**Interactions in Ultracold Gases**
From Atoms to Molecules

2003
ISBN 978-3-527-40389-9

# Optical Refrigeration

Science and Applications of Laser Cooling of Solids

*Edited by*
*Richard Epstein and Mansoor Sheik-Bahae*

WILEY-VCH Verlag GmbH & Co. KGaA

**The Editor**

**Richard Epstein**
Los Alamos National Lab.
Los Alamos, NM, USA
Epstein@lanl.gov

**Mansoor Sheik-Bahae**
University of New Mexico
Department of Physics and Astronomy
Albuquerque, NM, USA
msb@unb.edu

**Cover Illustration**
Printed with kind permission of the editors.

All books published by Wiley-VCH are carefully produced. Nevertheless, authors, editors, and publisher do not warrant the information contained in these books, including this book, to be free of errors. Readers are advised to keep in mind that statements, data, illustrations, procedural details or other items may inadvertently be inaccurate.

**Library of Congress Card No.:** applied for
**British Library Cataloguing-in-Publication Data:** A catalogue record for this book is available from the British Library.
**Bibliographic information published by the Deutsche Nationalbibliothek**
The Deutsche Nationalbibliothek lists this publication in the Deutsche Nationalbibliografie; detailed bibliographic data are available on the Internet at <http://dnb.d-nb.de>.

© 2009 WILEY-VCH Verlag GmbH & Co. KGaA, Weinheim

All rights reserved (including those of translation into other languages). No part of this book may be reproduced in any form by photoprinting, microfilm, or any other means nor transmitted or translated into a machine language without written permission from the publishers. Registered names, trademarks, etc. used in this book, even when not specifically marked as such, are not to be considered unprotected by law.

**Typesetting**   le-tex publishing services oHG, Leipzig
**Printing**   Strauss GmbH, Mörlenbach
**Binding**   Litges & Dopf GmbH, Heppenheim
**Cover Design**   Adam Design, Weinheim

Printed in the Federal Republic of Germany
Printed on acid-free paper

**ISBN**   978-3-527-40876-4

# Contents

Preface  IX

**1 Optical Refrigeration in Solids: Fundamentals and Overview**  1
*Richard I. Epstein and Mansoor Sheik-Bahae*
1.1  Basic Concepts  1
1.2  The Four-Level Model for Optical Refrigeration  4
1.3  Cooling Rare-Earth-Doped Solids  7
1.4  Prospects for Laser Cooling in Semiconductors  12
1.5  Experimental Work on Optical Refrigeration in Semiconductors  21
1.6  Future Outlook  26
    References  28

**2 Design and Fabrication of Rare-Earth-Doped Laser Cooling Materials**  33
*Markus P. Hehlen*
2.1  History of Laser Cooling Materials  33
2.2  Material Design Considerations  36
2.2.1  Active Ions  37
2.2.1.1  Rare-Earth Ions for Laser Cooling  37
2.2.1.2  Active Ion Concentration  39
2.2.2  Host Materials  40
2.2.2.1  Multiphonon Relaxation  40
2.2.2.2  Chemical Durability  42
2.2.2.3  Thermal and Thermomechanical Properties  42
2.2.2.4  Refractive Index  43
2.2.3  Material Purity  45
2.2.3.1  Vibrational Impurities  45
2.2.3.2  Metal-Ion Impurities  46
2.3  Preparation of High-Purity Precursors  48
2.3.1  Strategies for Preparing High-Purity Precursors  48
2.3.2  Process Conditions  50
2.3.2.1  Purity of Commercial Precursors  50

*Optical Refrigeration. Science and Applications of Laser Cooling of Solids.*
Edited by Richard Epstein and Mansoor Sheik-Bahae
Copyright © 2009 WILEY-VCH Verlag GmbH & Co. KGaA, Weinheim
ISBN: 978-3-527-40876-4

| | | |
|---|---|---|
| 2.3.2.2 | Process Equipment | 50 |
| 2.3.2.3 | Clean Environment | 51 |
| 2.3.3 | Material Purification | 51 |
| 2.3.3.1 | Filtration and Recrystallization | 51 |
| 2.3.3.2 | Solvent Extraction Using Chelating Agents | 52 |
| 2.3.3.3 | Fluorination and Drying in Hydrogen Fluoride Gas | 54 |
| 2.3.3.4 | Sublimation and Distillation | 55 |
| 2.3.3.5 | Electrochemical Purification | 57 |
| 2.3.4 | Determination of Trace Impurity Levels | 57 |
| 2.4 | Glass Fabrication | 59 |
| 2.4.1 | Glass Formation in $ZrF_4$ Systems | 59 |
| 2.4.2 | ZBLAN Glass Fabrication | 62 |
| 2.4.2.1 | Melting of the Starting Materials | 62 |
| 2.4.2.2 | Evaporative Losses | 63 |
| 2.4.2.3 | Dissolution and Homogenization | 63 |
| 2.4.2.4 | Optimum Rate of Cooling | 63 |
| 2.4.2.5 | Viscosity for Casting | 64 |
| 2.4.2.6 | Typical Glass Fabrication Parameters | 64 |
| 2.4.3 | Fluoride, Chloride, and Sulfide Glass Fabrication | 65 |
| 2.5 | Halide Crystal Growth | 65 |
| 2.6 | Promising Future Materials | 66 |
| 2.6.1 | Simplified Fluoride Glasses | 67 |
| 2.6.2 | Fluoride Crystals | 67 |
| 2.6.3 | Chloride and Bromide Crystals | 68 |
| | References | 68 |

| | | |
|---|---|---|
| **3** | **Laser Cooling in Fluoride Single Crystals** | **75** |
| | *Stefano Bigotta and Mauro Tonelli* | |
| 3.1 | Introduction | 75 |
| 3.2 | Physical Properties | 77 |
| 3.3 | Experimental | 78 |
| 3.3.1 | Growth Apparatus | 78 |
| 3.3.2 | Spectroscopic Setup | 80 |
| 3.3.3 | Cooling Setup | 81 |
| 3.4 | Spectroscopic Analysis | 83 |
| 3.5 | Cooling Results | 87 |
| 3.5.1 | Cooling Potential | 87 |
| 3.5.2 | Bulk Cooling | 89 |
| 3.6 | Conclusion | 93 |
| | References | 94 |

| | | |
|---|---|---|
| **4** | **$Er^{3+}$-Doped Materials for Solid-State Cooling** | **97** |
| | *Joaquin Fernandez, Angel Garcia-Adeva and Rolindes Balda* | |
| 4.1 | Low Phonon Energy Materials | 97 |
| 4.1.1 | $KPb_2Cl_5$ Crystal | 98 |
| 4.1.2 | Fluorochloride Glasses | 101 |

| | | |
|---|---|---|
| 4.2 | Internal Cooling Measurements   *101* | |
| 4.3 | Bulk Cooling Measurements   *105* | |
| 4.4 | Influence of Upconversion Processes on the Cooling Efficiency of $Er^{3+}$   *108* | |
| 4.4.1 | Spectroscopic Grounds: Upconversion Properties of $Er^{3+}$ Under Pumping in the $^4I_{9/2}$ Manifold   *108* | |
| 4.4.2 | A Phenomenological Cooling Model Including Upconversion   *111* | |
| | References   *114* | |
| | | |
| **5** | **Laser Refrigerator Design and Applications**   *117* | |
| | *Gary Mills and Mel Buchwald* | |
| 5.1 | Introduction   *117* | |
| 5.2 | Modeling   *119* | |
| 5.3 | Modeling Results   *121* | |
| 5.4 | Design Issues   *124* | |
| 5.5 | Mirror Heating   *129* | |
| 5.6 | Applications   *133* | |
| 5.6.1 | Comparison to Other Refrigeration Technologies   *133* | |
| 5.6.2 | Vibration   *133* | |
| 5.6.3 | Electromagnetic and Magnetic Noise   *134* | |
| 5.6.4 | Reliability and Lifetime   *134* | |
| 5.6.5 | Ruggedness   *134* | |
| 5.6.6 | Cryocooler Mass and Volume   *134* | |
| 5.6.7 | Efficiency and System Mass   *134* | |
| 5.6.8 | Cost   *136* | |
| 5.7 | Microcooling Applications   *136* | |
| | References   *138* | |
| | | |
| **6** | **Microscopic Theory of Luminescence and its Application to the Optical Refrigeration of Semiconductors**   *139* | |
| | *Greg Rupper, Nai H. Kwong and Rolf Binder* | |
| 6.1 | Introduction   *139* | |
| 6.2 | Microscopic Theory of Absorption and Luminescence   *141* | |
| 6.3 | Cooling Theory   *151* | |
| 6.4 | Cooling of Bulk GaAs   *153* | |
| 6.5 | Cooling of GaAs Quantum Wells   *159* | |
| 6.6 | Cooling of Doped Bulk Semiconductors   *162* | |
| 6.7 | Conclusion   *164* | |
| | References   *165* | |
| | | |
| **7** | **Improving the Efficiency of Laser Cooling of Semiconductors by Means of Bandgap Engineering in Electronic and Photonic Domains**   *169* | |
| | *Jacob B. Khurgin* | |
| 7.1 | Introduction   *169* | |

| 7.2 | Engineering the Density of States Using Donor–Acceptor Transitions  *171* |
|---|---|
| 7.3 | Refrigeration Using Phonon-Assisted Transitions  *174* |
| 7.4 | Laser Cooling Using Type II Quantum Wells  *180* |
| 7.5 | Photonic Bandgap for Laser Cooling  *186* |
| 7.6 | Novel Means of Laser Cooling Using Surface Plasmon Polaritons  *189* |
| 7.7 | Conclusions  *193* |
| | References  *194* |

**8 Thermodynamics of Optical Cooling of Bulk Matter**  *197*
*Carl E. Mungan*

| 8.1 | Introduction  *197* |
|---|---|
| 8.2 | Historical Review of Optical Cooling Thermodynamics  *198* |
| 8.3 | Quantitative Radiation Thermodynamics  *204* |
| 8.4 | Ideal and Actual Performance of Optical Refrigerators  *214* |
| 8.5 | Closing Remarks  *225* |
| | References  *230* |

**Index**  *233*

# Preface

Laser cooling of solids or "optical refrigeration" is a research area that encompasses basic scientific questions regarding the interaction of light with condensed matter systems and the practical issue of the design and construction of practical laser-powered cryocoolers. This volume brings together leading researchers to describe the critical issues being investigated and the approaches they are pursuing. There are two general thrusts to the current research programs: laser cooling of solids containing rare-earth (RE) ions, and the cooling of direct bandgap semiconductors. The advantages of using rare-earth-doped solids for laser cooling had been known for decades. The key optical transitions in RE-doped ions involve 4f electrons that are shielded by the filled 5s and 6s outer shells, which thus limit interactions with the surrounding lattice. Nonradiative decays due to multiphonon emission are thus suppressed.

Laser cooling of a solid was first experimentally demonstrated in 1995 with ytterbium-doped fluoride glass. Since then, researchers have demonstrated laser-induced cooling in a broad range of glasses and crystals doped with ytterbium. Later, laser cooling was achieved in thulium-doped fluoride glass. More recently, erbium-doped glasses and crystals have been laser cooled. Laser cooling of semiconductors has been more problematic. Direct bandgap semiconductors have several potential advantages over rare-earth-based cooling materials. These materials interact with light more strongly and have the potential to cool at much lower temperatures with higher cooling power densities. Additionally, they can be directly integrated into electronic and photonic devices. However, these materials have their own challenges. Semiconductors typically have high refractive indices that enhance total internal reflection and lead to luminescence trapping. Because of modest quantum efficiencies and heating from reabsorbed luminescence, net cooling is yet to be observed in semiconductors.

Chapter 1 provides an overview of research issues in optical refrigeration. It summarizes the physics of laser cooling in rare-earth-doped solids and in semiconductors and examines current research challenges. The factors that limit cooling performance in rare-earth-based coolers are parasitic heating from impurities and from components of the coolers such as the mirrors that form the cavity used to trap the pump radiation. The authors describe ongoing efforts to mitigate these problems. This chapter examines approaches to laser cooling in semiconductors

*Optical Refrigeration. Science and Applications of Laser Cooling of Solids.*
Edited by Richard Epstein and Mansoor Sheik-Bahae
Copyright © 2009 WILEY-VCH Verlag GmbH & Co. KGaA, Weinheim
ISBN: 978-3-527-40876-4

that attempt to overcome nonradiative recombination electron–hole excitations and luminescence trapping. In particular, it addresses the interplay of properties such as quantum efficiency, excitation density and extraction efficiency, which must be adjusted to achieve effective cooling.

In Chapter 2, Markus Hehlen describes the program at Los Alamos National Laboratory to develop ultrapure rare-earth-doped glasses that will be more efficient in laser cooling. He first summarizes the current status of the laser cooling of rare-earth-based laser cooling materials and then explains the advantages and constraints that must be considered when selecting rare-earth dopants and host materials. Dr. Hehlen discusses the harmful effects of impurities, such as OH ions and transition metals. He then describes the program in his laboratory to produce chemicals that have extremely low levels of impurities and to produce nearly defect-free glass samples for laser cooling. He ends with a discussion of halide crystal growth and some promising materials for laser cooling that may be developed in the future.

In Chapter 3, Bigotta and Tonelli discuss laser cooling in fluoride single crystals. They detail the growth of high-purity rare-earth-doped $BaY_2F_8$ (BYF) and $LiYF_4$ (YLF), and show experimental results from optical refrigeration in such crystals doped with Yb. Chapter 4 deals with optical refrigeration in erbium-doped materials. The authors Fernandez, Garcia-Adeva and Balda provide a detailed description of the synthesis of chloride crystal $KPb_2Cl_5$ and fluorochloride glass CNBZn doped with erbium. These hosts have very low maximum phonon energies, and thus have the potential to exhibit very high quantum efficiencies. This chapter also presents experimental data on the laser cooling of these Er-doped materials at $\lambda \approx 850$ nm. This is quite interesting, as it corresponds to a transition involving the ground state and the $^4I_{9/2}$ manifold, which is the second excited state. The authors also discuss the role of excited state absorption and upconversion in the laser cooling process of Er-doped systems.

Mills and Buchwald (Chapter 5) present a thorough analysis of the numerous practical issues involved in realizing an optical refrigerator based on rare-earth-doped materials. In particular, they consider a cooler based on Yb:ZBLAN and provide modeling results that include pump saturation, pump circulation in a nonresonant cavity, fluorescence shielding, and thermal link considerations.

Chapters 6 and 7 focus on laser cooling in semiconductors. Chapter 1 gave an overview of the macroscopic analysis and experimental issues in this field. Chapters 6 and 7 present rigorous theoretical calculations that illustrate the interplay of the physical processes involved in the laser cooling of semiconductors and to explore how the materials can be adjusted to optimize efficiency and operating temperatures. In Chapter 6, Rupper, Kwong and Binder present a comprehensive microscopic theory for absorption and luminescence in a direct-gap semiconductor under the pertinent condition of a partially ionized exciton gas. The authors also extend their analysis to 2D (quantum wells) as well as to doped bulk semiconductors. In Chapter 7, Khurgin addresses a number of important issues pertaining to laser cooling in semiconductors. In particular, he considers how to engineer the density of states, phonon-assisted absorption in the band-tail, cooling in type II quantum wells, photonic bandgap structures, and the escape of luminescence by means of surface plasmon coupling.

In Chapter 8, Mungan reviews the concepts and history of the thermodynamics of fluorescent cooling. He discusses how the entropy and energy of beams of radiation are related. Mungan uses these results to calculate the ideal coefficient of performance for laser cooling and discusses how real-world effects reduce the efficiency. He then examines important practical topics such as radiation-balanced lasing, in which the heat generated by lasing is compensated for by fluorescent cooling, and the recycling of output optical energy into the input in order to increase the cooling efficiency.

In short, we hope that this book will serve its purpose as a major collection of the most significant findings to date in this relatively young yet thriving area of research. While the authors in this volume have attempted to provide forward-looking accounts of their areas of research, we have no doubt that the rapid and continuous progress in optical refrigeration will soon necessitate a new compendium of progress in the laser cooling of various solids and the development of optical refrigeration devices.

Albuquerque, NM  
October 2008

Mansoor Sheik Bahae  
Richard Epstein

# List of Contributors

**Stefano Bigotta**
INFN
Sezione di Pisa
NEST-CNR
Università di Pisa
Dipartimento di Fisica
Largo Bruno Pontecorvo 3
56127 Pisa
Italy

**Richard Epstein**
Los Alamos National Laboratory
MS D466
Los Alamos, NM 87545
USA

**Markus P. Hehlen**
Los Alamos National Laboratoy
Mail Stop J565
Los Alamos, NM 87545
USA

**Gary Mills**
Ball Aerospace & Technologies Corp.
1600 Commerce Street
Boulder, CO 80301
USA

**Mauro Tonelli**
INFN
Sezione di Pisa
NEST-CNR
Università di Pisa
Dipartimento di Fisica
Largo Bruno Pontecorvo 3
56127 Pisa
Italy

**Joaquin Fernandez**
University of the Basque Country
Applied Physics Department I
Escuela Superior de Ingenieros
Alda. de Urquijo s/n
48013 Bilbao
Spain

**Carl E. Mungan**
United States Naval Academy
Physics Department
572C Holloway Road
Annapolis, MD 21402-5002
USA

**Rolf Binder**
University of Arizona
College of Optical Sciences/Department of Physics
1630 East University Boulevard
Tucson, AZ 85721
USA

*Jacob Khurgin*
Johns Hopkins University
Department of Electrical and Computer Engineering
315 Barton Hall
3400 N. Charles Street
Baltimore, MD 21218
USA

*Mansoor Sheik-Bahae*
University of New Mexico
Department of Physics and Astronomy
800 Yale Blvd. MSC07 4220
Albuquerque, NM, 87131
USA

*Angel Garcia-Adeva*
University of the Basque Country
Applied Physics Department I
Escuela Superior de Ingenieros
Alda. de Urquijo s/n
48013 Bilbao
Spain

*Rolindes Balda*
University of the Basque Country
Applied Physics Department I
Escuela Superior de Ingenieros
Alda. de Urquijo s/n
48013 Bilbao
Spain

*Mel Buchwald*
Buchwald Consulting
Santa Fe, NM 87501
USA

*Greg Rupper*
University of Arizona
College of Optical Sciences
1630 East University Boulevard
Tucson, AZ 85721
USA

*Nai H. Kwong*
University of Arizona
College of Optical Sciences
1630 East University Boulevard
Tucson, AZ 85721
USA

# Acknowledgments

We wish to thank our research groups (past and present) at the University of New Mexico and Los Alamos National Laboratory, as well as our numerous colleagues who have been instrumental in the progress of the research that has culminated to the publication of this book.

We also thank NASA, the US Department of Energy, and the Air Force Office of Scientific Research (AFOSR) for supporting optical refrigeration research. We are especially grateful to Dr. Charlie Stein (Air Force Research Lab) and Dr. Kent Miller (at AFOSR) for their continued and enthusiastic support of this research program.

# 1
## Optical Refrigeration in Solids: Fundamentals and Overview
*Richard I. Epstein and Mansoor Sheik-Bahae*

### 1.1
### Basic Concepts

Optical refrigeration, or the cooling of solids with near-monochromatic light, is a discipline that was anticipated almost 80 years ago, decades before the invention of the laser. In 1929, the German physicist Peter Pringsheim (Figure 1.1) proposed the cooling of solids by fluorescence upconversion [3]. Because of the great advances that have been made in the use of lasers to cool and trap dilute gases of atoms and ions to extremely low temperatures, the term "laser cooling" is most often used in reference to this area of science. This is quite justified given its spectacular achievements, such as the creation of Bose–Einstein condensates and many related phenomena [1, 2]. Nevertheless, optical refrigeration is itself a rapidly growing field; one that offers insights into the interaction of light with condensed matter, and has the potential to provide the basis for new types of cryogenic refrigeration. In the solid phase, thermal energy is largely contained in the vibrational modes of the lattice. In the laser cooling of solids, light quanta in the red tail of the absorption spectrum are absorbed from a monochromatic source, and then spontaneous emission of more energetic (blue-shifted) photons occurs. The extra energy is extracted from lattice phonons, the quanta of vibrational energy that are generated from heat. The removal of these phonons is therefore equivalent to cooling the solid. This process has also been termed "anti-Stokes fluorescence" and "luminescence upconversion" cooling.

Laser cooling of solids can be exploited to achieve an all-solid-state cryocooler [4–6], as conceptually depicted in Figure 1.2. The advantages of compactness, no vibrations, no moving parts or fluids, high reliability, and no need for cryogenic fluids have motivated intensive research. Spaceborne infrared sensors are likely to be the first beneficiaries, with other applications requiring compact cryocooling reaping the benefits as the technology progresses. A study by Ball Aerospace Corporation [7] shows that in low-power spaceborne operations, ytterbium-based optical refrigeration could outperform conventional thermoelectric and mechanical coolers in the temperature range 80–170 K. Efficient, compact semiconductor lasers can pump optical refrigerators. In many potential applications, the requirements

*Optical Refrigeration. Science and Applications of Laser Cooling of Solids.*
Edited by Richard Epstein and Mansoor Sheik-Bahae
Copyright © 2009 WILEY-VCH Verlag GmbH & Co. KGaA, Weinheim
ISBN: 978-3-527-40876-4

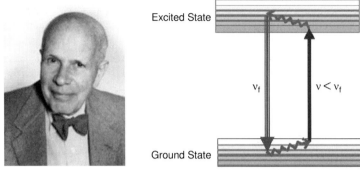

**Figure 1.1** In 1929, Peter Pringsheim suggested that solids could cool through anti-Stokes fluorescence, in which a substance absorbs a photon and then emits one of greater energy. The energy diagram on the *right* shows one way in which this could occur. An atom with two broad levels is embedded in a transparent solid. The light source of frequency $h\nu$ excites atoms near the top of the ground state level to the bottom of the excited state. Radiative decays occurring after thermalization emit photons with average energy $h\nu_f > h\nu$.

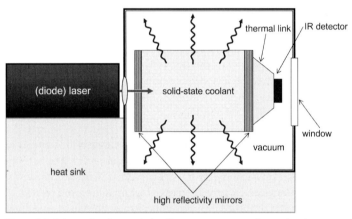

**Figure 1.2** Schematic of an optical refrigeration system. Pump light is efficiently generated by a semiconductor diode laser. The laser light enters the cooler through a pinhole in one mirror and is trapped by the mirrors until it is absorbed. Isotropic fluorescence escapes the cooler element and is absorbed by the vacuum casing. A sensor or some other load is connected in the *shadow region* of the second mirror. Figure 1.2 has been reproduced from [6].

on the pump lasers are not very restrictive. The spectral width of the pump light has to be narrow compared to the thermal spread of the fluorescence. A multimode fiber-coupled laser with a spectral width of several nanometers would be adequate. In an optical refrigerator, the cooling power is of the order 1% of the pump laser power. Only modest lasers are adequate for microcooling applications with a heat lift of mW. For larger heat lifts, correspondingly more powerful lasers are needed. In all cooling applications, the cooling element has to be connected to the device being cooled, the *load*, by a thermal link; see Figure 1.2. This link siphons heat from the load while preventing the waste fluorescence from hitting the load and heating it. See also Chapter 5 by Mills and Buchwald for further discussion on the practical implementation of all-solid-state cryocoolers in rare-earth-doped solids.

Another potential application of laser cooling of solids is to eliminate heat production in high-power lasers. Even though laser emission is always accompanied by heat production, Bowman [8, 9] realized that in some laser materials, the pump wavelength can be adjusted so that the spontaneous anti-Stokes fluorescence cooling compensates for the laser heating. Such a thermally balanced laser would not suffer from thermal defocusing or heat damage.

The process of optical refrigeration can occur only in special high-purity materials (see Section 1.3) that have appropriately spaced energy levels and emit light with a high quantum efficiency. To date, optical refrigeration research has been confined to glasses and crystals doped with rare-earth elements and direct-band semiconductors such as gallium arsenide. Laser cooling of rare-earth-doped solids has been successfully demonstrated, while observations of net cooling in semiconductors have remained elusive. Figure 1.1 schematically depicts the optical refrigeration processes for a two-level system with vibrationally broadened ground- and excited-state manifolds. Photons from a low-entropy light source (i.e. a laser) with energy $h\nu$ excite atoms from the top of the ground state to the bottom of the excited state. The excited atoms reach quasi-equilibrium with the lattice by absorbing phonons. Spontaneous emission (fluorescence) follows, with a mean photon energy $h\nu_f$ that is higher than that of the absorbed photon. This process has also been called anti-Stokes fluorescence. There were initial concerns that the second law of thermodynamics might be violated until Landau clarified the issue in 1946 by assigning an entropy to the radiation [10].

In the aforementioned simple model, the interaction rate between electrons and phonons within each manifold is assumed to be far faster than the spontaneous emission rate, which is valid for a broad range of materials and temperatures. The cooling efficiency or fractional cooling energy for each photon absorbed is

$$\eta_c = \frac{h\nu_f - h\nu}{h\nu} = \frac{\lambda}{\lambda_f} - 1, \qquad (1.1a)$$

where $\lambda = c/\nu$ is the wavelength. The invention of the laser in 1960 prompted several unsuccessful attempts to observe laser cooling of solids [11–13]. In 1995, net cooling was first achieved by workers at Los Alamos National Laboratory [14]. Two technical challenges were addressed and overcome in these experiments. The Los Alamos researchers had to have a system in which (i) the vast majority of optical excitations recombine radiatively and (ii) there is a minimal amount of parasitic heating due to unwanted impurities. Both of these critical engineering issues are ignored in the idealized situation described by (1.1a), but are key to experimental success.

It is also important that spontaneously emitted photons escape the cooling material without being trapped and re-absorbed, which would effectively inhibit spontaneous emission [15, 16]. This is a critical issue for high-index semiconductors where total internal reflection can cause strong radiation trapping. In the absence of radiation trapping, the fraction of atoms that decay to the ground state by the desired radiative process is known as the quantum efficiency, $\eta_q = W_{rad}/(W_{rad} + W_{nr})$, where $W_{rad}$ and $W_{nr}$ are radiative and nonradiative decay rates, respectively. Including a fluorescence escape efficiency $\eta_e$ defines an external quantum efficiency (EQE), $\eta_{ext} = \eta_e W_{rad}/(\eta_e W_{rad} + W_{nr})$, which assumes the fluorescence is reabsorbed within the excitation volume (see Section 1.4). This describes the efficiency by which a photoexcited atom decays into an escaped fluorescence photon in free space. In a similar fashion, an absorption efficiency $\eta_{abs} = \alpha_r/(\alpha_r + \alpha_b)$ is defined to account for the fraction of pump laser photons that are engaged in cooling [6, 17]. Here $\alpha_r$ is the resonant absorption coefficient and $\alpha_b$ is the unwanted parasitic (background) absorption coefficient. As will be derived in Sections 1.2 and 1.4, the combination of all of these effects redefines the cooling efficiency as:

$$\eta_c = \eta_{ext}\eta_{abs}\frac{\lambda}{\lambda_f} - 1, \tag{1.1b}$$

where the product $\eta_{ext}\eta_{abs}$ indicates the efficiency of converting an absorbed laser photon to an escaped fluorescence photon. Note that $\eta_{abs}$ is frequency dependent and falls off rapidly below a photon energy $h\nu_f - k_B T$, where $k_B$ is the Boltzmann constant and $T$ is the lattice temperature. At pump photon energies of much more than $k_B T$ below $h\nu_f$, $\eta_{abs}$ is too small to make $\eta_c > 0$ and laser cooling is unattainable. The above analysis defines the approximate condition needed for laser cooling [6, 17]:

$$\eta_{ext}\eta_{abs} > 1 - \frac{k_B T}{h\nu_f}. \tag{1.2}$$

This relation quantifies the required efficiencies: cooling a material from room temperature with a nominal energy gap (pump photon) of 1 eV from room temperature demands that $\eta_{ext}\eta_{abs} > 97\%$. Although suitable lasers were available in the early 1960s, more than three decades of progress in material growth were needed to satisfy this condition.

## 1.2
### The Four-Level Model for Optical Refrigeration

Consider the four-level system of Figure 1.3 in which the ground-state manifold consist of two closely spaced levels of $|0\rangle$ and $|1\rangle$ with an energy separation of $\delta E_g$. The excited manifold consists of two states $|3\rangle$ and $|2\rangle$ with an energy separation $\delta E_u$. Laser excitation at $h\nu$ is tuned to be in resonance with the $|1\rangle - |2\rangle$ transition, as shown by the solid red arrow. The double-line arrows depict the spontaneous emission transitions from the upper level to the ground states with a rate of $W_{rad}$; this rate is assumed to be the same for all four transitions. The nonradiative decay rates (indicated by the dotted lines) are also assumed to be equal and given by $W_{nr}$. The population in each manifold reaches a quasi-thermal equilibrium via an

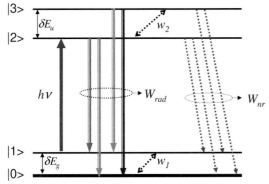

**Figure 1.3** The four-level energy model for optical refrigeration consisting of two pairs of closely spaced levels: $|0\rangle$ and $|1\rangle$ in the ground state and $|2\rangle$ and $|3\rangle$ in the excited-state manifolds.

electron–phonon interaction rate given by $w_1$ and $w_2$ for lower and upper states, respectively.

The rate equations governing the density populations $N_0$, $N_1$, $N_2$, and $N_3$ are:

$$\frac{dN_1}{dt} = -\sigma_{12}\left(N_1 - \frac{g_1}{g_2}N_2\right)\frac{I}{h\nu} + \frac{R}{2}(N_2 + N_3) - w_1\left(N_1 - \frac{g_1}{g_0}N_0 e^{-\delta E_g/k_B T}\right), \tag{1.3a}$$

$$\frac{dN_2}{dt} = \sigma_{12}\left(N_1 - \frac{g_1}{g_2}N_2\right)\frac{I}{h\nu} - RN_2 + w_2\left(N_3 - \frac{g_3}{g_2}N_2 e^{-\delta E_u/k_B T}\right), \tag{1.3b}$$

$$\frac{dN_3}{dt} = -RN_3 - w_2\left(N_3 - \frac{g_3}{g_2}N_2 e^{-\delta E_u/k_B T}\right), \tag{1.3c}$$

$$N_0 + N_1 + N_2 + N_3 = N_t, \tag{1.3d}$$

where $R = 2W_{rad} + 2W_{nr}$ is the total upper state decay rate, $\sigma_{12}$ is the absorption cross-section associated with the $|1\rangle$–$|2\rangle$ transition, $I$ is the incident laser irradiance, and the $g_i$ terms represent degeneracy factors for each level. The weighting factor in the electron–phonon interaction terms ($w_1$ and $w_2$) maintains the Boltzmann distribution among each manifold at quasi-equilibrium. The net power density deposited in the system is the difference between the absorbed and the radiated contributions:

$$P_{net} = \sigma_{12}N_1 1 - \frac{g_1 N_2}{g_2 N_1} I - W_{rad}[N_2(E_{21} + E_{20}) + N_3(E_{31} + E_{30})] + \alpha_b I, \tag{1.4}$$

where the first term is the laser excitation ($|1\rangle$–$|2\rangle$ transition) and the second term includes the spontaneous emission terms from levels $|2\rangle$ and $|3\rangle$ with their respective photon energies. We have also included a term that represents the parasitic absorption of the pump laser with an absorption coefficient of $\alpha_b$. It is straightforward to evaluate the steady-state solution to the above rate equations by setting the time derivatives to zero. To emphasize certain features, we ignore saturation and assume a degeneracy of unity for all levels. The net power density is then obtained as:

$$P_{net} = \alpha I \left(1 - \eta_q \frac{h\nu_f}{h\nu}\right) + \alpha_b I, \tag{1.5}$$

where $\eta_q = (1 + W_{nr}/W_{rad})^{-1}$ is the (internal) quantum efficiency and $h\nu_f$ denotes the mean fluorescence energy of the four-level system given by:

$$h\nu_f = h\nu + \frac{\delta E_g}{2} + \frac{\delta E_u}{1 + (1 + R/w_2)e^{\delta E_u/k_B T}}. \tag{1.6}$$

The ground-state resonant absorption $\alpha$ is given by:

$$\alpha = \sigma_{12} N_t \left(1 + e^{\delta E_g/k_B T}\right)^{-1}. \tag{1.7}$$

Despite its simplicity, the four-level model reveals essential features of solid-state optical refrigeration. First, (1.7) exhibits diminishing pump absorption due to thermal depletion of the top ground state at low temperatures, $k_B T < \delta E_g$. This implies that the width of the ground-state manifold ($\delta E_g$) must be narrow to achieve cooling at low temperatures with reasonable efficiency. This issue will be revisited when discussing semiconductors in Section 1.4. Second, (1.6) shows that the mean fluorescence photon energy is redshifted at low temperatures, which further lowers the cooling efficiency. This shift is enhanced if the electron–phonon interaction rate ($w_2$) is smaller than the upper state recombination rate ($R$). This means that if $w_2 < R$, the excited state can decay before thermalization with the lattice, which results in no fluorescence upconversion and no cooling [18]. This extreme limit of cold electron recombination is an issue for semiconductors at very low temperatures, where the electrons primarily interact with the lattice via relatively slow acoustic phonons [15].

Dividing (1.5) by the total absorbed power density $P_{abs} = (\alpha + \alpha_b)I$ gives the cooling efficiency $\eta_c = -P_{net}/P_{abs}$:

$$\eta_c = \eta_q \eta_{abs} \frac{h\nu_f}{h\nu} - 1, \tag{1.8}$$

which is similar to (1.1b), not including the luminescence trapping. The most useful feature of the four-level model is its description of the temperature dependence of the cooling in a physically transparent manner. As the temperature is lowered, the redshifting of the mean fluorescence wavelength and the reduction of the resonant absorption reduce the cooling efficiency. At the temperature $T = T_m$, the cooling stops (i.e. $\eta_c (T_m) = 0$). This minimum achievable temperature ($T_m$) can be lowered by reducing the background absorption (higher purity), increasing the quantum efficiency, and enhancing the resonant absorption (e.g. choosing a material with a narrow ground-state manifold). The effect of fluorescence trapping and its consequent reabsorption by both resonant and parasitic processes will further diminish the quantum efficiency. We will discuss this in detail when we analyze laser cooling in semiconductors where total internal reflection leads to substantial trapping.

## 1.3
## Cooling Rare-Earth-Doped Solids

The advantages of rare-earth (RE)-doped solids for laser cooling had been foreseen for decades. Kastler (1950) [11] and Yatsiv (1961) [13] suggested that these materials could be used for optical cooling. The key optical transitions in RE-doped ions involve 4f electrons that are shielded by the filled 5s and 6s outer shells, which limit interactions with the surrounding lattice. Nonradiative decays due to multiphonon emission are thus suppressed. Hosts with low phonon energy (e.g. fluoride crystals and glasses) further diminish nonradiative decay and hence boost quantum efficiency. In 1968, Kushida and Geusic [12] attempted to cool a $Nd^{3+}$:YAG crystal with 1064 nm laser radiation. They reported a reduction in heating but no cooling; it is unclear whether they observed any anti-Stokes cooling effects. Laser cooling of a solid was first experimentally demonstrated in 1995 with the ytterbium-doped fluorozirconate glass ZBLANP:$Yb^{3+}$ [14]. Laser-induced cooling has since been observed in a range of glasses and crystals doped with $Yb^{3+}$ (ZBLANP [19–22], ZBLAN [23, 24], CNBZn [9, 25] BIG [25, 26], $KGd(WO_4)_2$ [9], $KY(WO_4)_2$ [9], YAG [27], $Y_2SiO_5$ [27], $KPb_2Cl_5$ [25,28], $BaY_2F_8$ [29–31], and YLF [32, 33]). See also Chapters 2–5 for further reading on this subject, ranging from synthesis to practical implementations of optical refrigeration in various rare-earth-doped solids.

Figure 1.4 shows the cooling and heating of a sample of $Yb^{3+}$-doped ZBLANP for a range of pump wavelengths. The data in Figure 1.4 were obtained using a setup similar to that of Figure 1.2, where the pump laser is circulated in a nonresonant cavity by bouncing off dielectric mirrors deposited on the sample [5]. For wave-

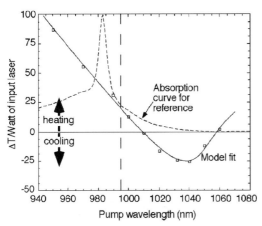

**Figure 1.4** The temperature change (normalized to the incident power) in ytterbium-doped ZBLANP glass as a function of pump wavelength. When the pump wavelength is considerably longer than the mean wavelength of the fluorescence $\lambda_F$ (vertical dashed line), the escaping light carries more energy than the absorbed laser light and the glass cools. Heating at wavelengths greater than $\lambda_F$ is due to the imperfect quantum efficiency of the fluorescence and nonresonant light absorption [5].

lengths that are shorter than the mean fluorescence wavelength $\lambda_F$ (vertical dashed line), the sample heats up due to the Stokes shift as well as nonradiative processes. At longer wavelengths, anti-Stokes cooling dominates, and cooling as large as 25 K/W of absorbed laser power is measured. At still longer wavelengths, absorption by impurities or imperfections dominates, and the sample heats.

In 2000, laser cooling in $Tm^{3+}$-doped ZBLANP was reported at $\lambda \sim 1.9$ µm [34]. The significance of this result was twofold. First, it verified the scaling law of (1.1a) and (1.1b) by demonstrating that there was a factor of almost two enhancement in the cooling efficiency compared to Yb-doped systems. The enhancement scales as the ratio of the corresponding cooling transition wavelengths. Second, it was the first demonstration of laser cooling in the presence of excited state absorption. A record cooling power of $\sim 73$ mW was obtained in this material by employing a multipass geometry [35]. More recently, cooling of $Er^{3+}$-doped glass (CNBZn) and crystal ($KPb_2Cl_5$) at $\lambda \sim 0.870$ µm was reported by a Spanish group ([36], see also Chapter 4 for more details). It is interesting to note that the cooling transition used in these experiments is between the ground state and the fourth excited state ($^4I_{9/2}$) of $Er^{3+}$, not the first excited state as illustrated in Figure 1.1. High-energy transitions have lower cooling efficiencies (1.1) but potentially higher quantum efficiencies due to their low nonradiative decay rates to the ground state. The presence of higher excited states in $Er^{3+}$ may prove advantageous, since the energy upconversion transitions (i.e. at the cooling wavelengths of the main transition) are endothermic, with a high quantum efficiency [36, 37]. This is also the case with the cooling of $Tm^{3+}$ [34].

The initial proof-of-principle experiments in ZBLANP:$Yb^{3+}$ achieved cooling by an amount of 0.3 K below ambient temperature [14]. The LANL group has since cooled ZBLANP:$Yb^{3+}$ to 208 K starting from room temperature [22], as shown in Figure 1.5. Although progress is being made, optical refrigerators need to be more efficient and operate at lower temperatures, below about 170 K, to be competitive with other solid-state coolers such as thermoelectric (Peltier) devices. Several studies have shown that ytterbium- or thulium-doped solids should be capable of providing efficient cooling at temperatures well below 100 K [4, 27, 38].

There are several factors that limit the cooling of rare-earth-doped solids in available materials. The most significant factor is the choice of laser-cooling medium. The ideal cooling efficiency (1.1) shows that there is an advantage of pumping with lower-energy photons. This increased efficiency was part of the motivation for investigating thulium-doped cooling materials, since their ground- and excited-state manifolds are separated by about 0.6 eV, compared to 1.2 eV in ytterbium-doped solids. There are obstacles, however, when moving to longer wavelengths. The first is the limited choice in relation to pump lasers, since there are fewer available near 0.6 eV than near 1.2 eV. While this is not a fundamental consideration, it needs to considered for near-term commercialization. A second and more general reason involves the ratio of radiative to nonradiative relaxation decays. The rate of nonradiative, heat-producing, multiphonon decay decreases exponentially with the separation between the ground- and first excited-state manifolds; this is the well-known energy-gap law. In practical terms, this means that because of the relatively large energy of the excited level in ytterbium-doped materials, nonradiative decays do not significantly decrease the quantum efficiency. For pure thulium-doped material, nonradiative decay can overwhelm anti-Stokes cooling, depending on the properties of the host material. For materials with low maximum phonon energies, such as ZBLANP and other fluoride hosts, the nonradiative decays are relatively slow. Many thulium-doped oxide crystals and glasses have rapid nonradiative decay rates that prevent laser cooling.

Another consideration in the choice of cooling medium is the width of the ground-state manifold. According to Boltzmann statistics, lower energy levels in the manifold are more populated than higher ones. As the temperature falls and $k_B T$ becomes small compared to the energy width of the ground-state manifold,

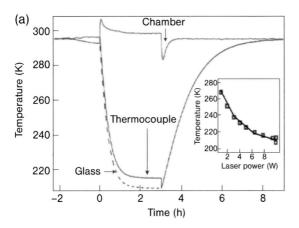

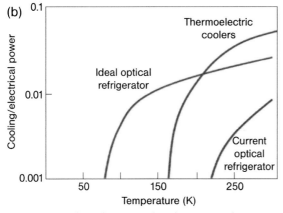

**Figure 1.5** Panel (a) shows record cooling to 208 K with ZBLANP:Yb$^{3+}$. The temperatures are measured with thermocouples on the sample and chamber; the internal temperature of the glass is inferred [22]. Panel (b) compares the cooling efficiencies of available thermoelectric coolers (TECs) with ZBLANP:Yb$^{3+}$-based optical refrigerators. Devices based on materials with low parasitic heating will outperform TECs below 200. Coolers made from current materials are less efficient than TECs at all temperatures [39]. Figure 1.5a has been reproduced from [22].

the upper levels become depopulated, leading to increased transparency at lower frequencies; this effect is illustrated in the four-level system discussed above. The net effect is that, at low temperatures, the numerator of (1.2) becomes small and cooling efficiency goes to zero; see (1.1b). The width of the ground-state manifold is typically the result of crystal field splitting and depends on both the dopant ion and the host material. By choosing ions and a host that give narrow ground-state manifolds, the material can cool to lower temperatures before the low-frequency transparency condition sets in.

For the material systems studied so far, cooling is not limited by the reasons outlined above. It is most likely hindered by parasitic heating in the bulk of the cooling material or on its surface. As one can see in Figure 1.5b, the cooling efficiencies of currently available ZBLANP:$Yb^{3+}$ are far below that of an ideal material with no parasitic heating. One important source of heating in this material is the quenching of excited ytterbium ions by impurities such as iron and copper. The radiative decay time of an excited $Yb^{3+}$ ion is about 1 ms. During this time, the excitation migrates through the glass by transferring energy to neighboring ions. If the excitation encounters an impurity atom, the energy can be transferred to this atom and rapidly converted into heat. A detailed study by Hehlen et al. [39] found that the ideal cooling efficiency can be approached when the concentration of an impurity such as $Cu^{2+}$ is less than 0.01 ppm and that for $Fe^{2+}$ is below 0.1 ppm; see Figure 1.6. See also Chapter 2 by Hehlen.

An additional source of parasitic heating is absorption in the mirrors that trap the pump radiation in the cooling element. In the LANL experiments, the cooling glass has a pair of high-reflectivity mirrors deposited on two surfaces, as depicted in Fig-

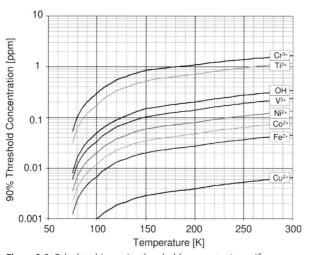

**Figure 1.6** Calculated impurity threshold concentrations. If the impurity level of an ion is above the level shown here, the cooling efficiency of the ZBLAN:1%$Yb^{3+}$ will be less than 90% of its ideal value and rapid heat conversion can occur. See the detailed study by Hehlen et al. [39].

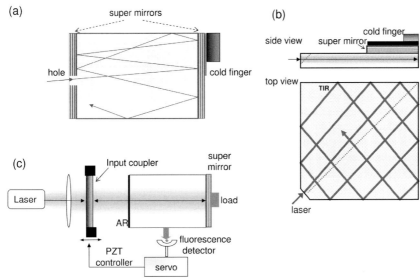

**Figure 1.7** Methods to enhance pump absorption. (a) Non-resonant cavity formed by dielectric mirrors deposited on the sample. The pump laser enters through a small hole in one of the mirrors. (b) Nonresonant cavity based on total internal reflection, in which the laser enters through a bevel. (c) Cavity-enhanced absorption, where light couples through a matched dielectric mirror. This required active stabilization.

ures 1.2 and 1.7a. Pump light is reflected multiple times by each mirror, so that that even a relatively low absorption of 0.0001 per surface produces significant heating. The deposition of higher-quality dielectric mirrors, which is currently undertaken by LANL/UNM teams, should obviate this problem. An alternative approach is to avoid dielectric mirrors altogether and exploit total internal refection to circulate the pump beam, as depicted in Figure 1.7b [40]. A proof-of-concept experiment by UNM demonstrated more than ten roundtrips and $\Delta T \approx 8$ K in Tm:ZBLANP [40]. However, the absorbed power was limited by the imperfections in the right-angle corners, which were found to be difficult to control due to the mechanical properties of ZBLANP. In Chapter 5, Mills and Buchwald discuss another variation of this technique recently investigated by Ball Aerospace Inc. Another method of enhancing pump absorption is to use resonant cavity effects. Both intra-laser-cavity [24] and external resonant cavity [41] geometries have been demonstrated. The latter approach, as depicted in Figure 1.7c, has been found to be capable of achieving pump absorption exceeding 90% [41]. Most recently, this method was employed using active stabilization to successfully achieve $\Delta T \approx 70$ K in a Yb:YLF crystal [33]. This is a highly promising result considering that it was obtained with a full blackbody thermal load that is nearly five times higher than that reported in [22]. It has also been proposed that photon localization in nanocrystalline powders can be exploited to enhance laser pump absorption in the cooling of rare-earth-doped systems [42].

## 1.4
### Prospects for Laser Cooling in Semiconductors

Researchers have examined other condensed matter systems beyond RE-doped materials, with an emphasis on semiconductors [17, 43–46]. Semiconductor coolers provide more efficient pump light absorption, the potential for much lower temperatures, and the opportunity for direct integration into electronic and photonic devices. However, these materials provide their own set of engineering challenges, and net cooling is yet to be observed. The essential difference between semiconductors and RE-doped materials is in their cooling cycles. In the latter, the cooling transition occurs in localized donor ions within the host material, while the former involves transitions between extended valence and conduction bands of a direct gap semiconductor (see Figure 1.8a). Indistinguishable charge carriers in Fermi–Dirac distributions may allow semiconductors to get much colder than RE materials. The highest energy levels of the ground state manifold in the RE-doped systems become less populated as the temperature is lowered, due to Boltzmann statistics. The cooling cycle becomes ineffective when the Boltzmann constant times the lattice temperature becomes comparable to the width of the ground state (see previous section describing the four-level model). This sets a limit of $T \sim 100$ K for most existing RE-doped systems. No such limitation exists in pure (undoped) semiconductors – temperatures as low as 10 K may be achievable [15, 17, 47]. See also Chapter 6 by Rupper, Kwong and Binder.

Semiconductors should achieve a higher cooling power density than RE materials. The maximum cooling power density (rate of heat removal) is $\approx N \times k_B T/\tau_r$, where $N$ is the photoexcited electron (hole) density and $\tau_r$ is the radiative recombination time. In semiconductors, the optimal density $N$ is limited due to many-body processes and does not exceed that of moderately doped RE systems. We can gain

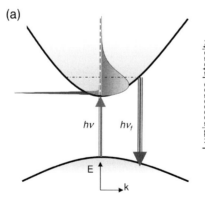
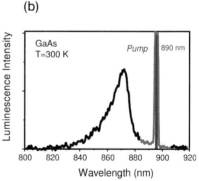
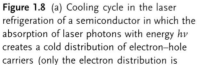

**Figure 1.8** (a) Cooling cycle in the laser refrigeration of a semiconductor in which the absorption of laser photons with energy $h\nu$ creates a cold distribution of electron–hole carriers (only the electron distribution is shown for clarity). The carriers then heat up by absorbing phonons, and this is followed by an upconverted luminescence at $h\nu_F$. (b) Typical anti-Stokes luminescence observed in a GaAs/GaInP double heterostructure [6].

5–6 orders of magnitude in cooling power density because the radiative recombination rates in semiconductors are much faster than in RE ions.

Laser cooling of semiconductors has been examined theoretically [15, 44, 45, 47–52], as well as in experimental studies [46, 53–56]. A feasibility study by the authors outlined the conditions needed for net cooling based on fundamental material properties and light management [15]. Researchers at the University of Arizona ( [47, 50], Chapter 6) studied luminescence upconversion in the presence of partially ionized excitons, which are understood to attain temperatures approaching 10 K. The role of bandtail states [52] and the possible enhancement of laser cooling by including the effects of photon density of states as well as novel luminescence coupling schemes based on surface plasmon polaritons [57, 58] were recently introduced by Khurgin at Johns Hopkins University. Further details on these issues can be found in Chapter 7 by Khurgin. Here, we expand on the basic model of [15] and present the theoretical foundation of laser cooling in semiconductor structures with an arbitrary external efficiency. This treatment accounts for the luminescence redshift due to reabsorption, the effect of the parasitic absorption of the pump, the luminescence power, and band-blocking effects. We then discuss the latest experimental results from attempts to achieve the first observations of laser cooling in a semiconductor material.

We consider an intrinsic (undoped) semiconductor system uniformly irradiated with laser light at photon energy $h\nu$. Furthermore, we assume that only a fraction $\eta_e$ of the total luminescence can escape the material while the remaining fraction $(1 - \eta_e)$ is trapped and recycled, thus contributing to carrier generation. For now, we will ignore the parasitic absorption of luminescence, but we will consider its implications later. For a given temperature, the rate equation for the electron–hole pair density ($N$) is given by [15]:

$$\frac{dN}{dt} = \frac{\alpha I}{h\nu} - AN - BN^2 - CN^3 + (1 - \eta_e)BN^2 . \tag{1.9}$$

Here $\alpha(\nu, N)$ is the interband absorption coefficient, which includes many-body and blocking factors. The recombination process consists of nonradiative ($AN$), radiative ($BN^2$), and Auger ($CN^3$) rates. All of the above coefficients are temperature dependent. The last term represents the increase in $N$ from the reabsorption of the luminescence that does not escape, assuming that the reabsorption occurs within the laser excitation volume. The density dependence of $\alpha$ results from both Coulomb screening and band-blocking (saturation) effects. The latter can be approximated by a blocking factor such that [59, 60]:

$$\alpha(N, h\nu) = \alpha_0(N, h\nu)\{f_v - f_c\} , \tag{1.10}$$

where $\alpha_0$ denotes the unsaturated absorption coefficient. The strongly density-dependent blocking factor in the parentheses [61] contains Fermi–Dirac distribution functions for the valence ($f_v$) and conduction ($f_c$) bands.

Under steady-state conditions, (1.9) can be rewritten as

$$0 = \frac{\alpha(\nu, N)}{h\nu} I - AN - \eta_e BN^2 - CN^3 . \tag{1.11}$$

This indicates that the fluorescence trapping effectively inhibits the spontaneous emission as it appears through $\eta_e B$ only. This result has also been shown previously by Asbeck [16]. It is important to note that $\eta_e$ is itself an averaged quantity over the entire luminescence spectrum.

$$\eta_e = \frac{\int S(\nu) R(\nu) \, d\nu}{\int R(\nu) \, d\nu} . \tag{1.12}$$

Here $S(\nu)$ is the geometry-dependent escape probability of photons with energy $h\nu$, and $R(\nu)$ is the luminescence spectral density, which is related to the absorption coefficient through reciprocity using a "nonequilibrium" van Roosbroeck–Shockley relation (also known as the Kubo–Martin–Schwinger (KMS) relation) [59, 62]:

$$R(\nu, N) = \frac{8\pi n^2 \nu^2}{c^2} \alpha(\nu, N) \left\{ \frac{f_c(1 - f_v)}{f_v - f_c} \right\}, \tag{1.13}$$

where $c$ is the speed of light and $n$ is the index of refraction. Note that the radiative recombination coefficient $B$ is obtained by $BN^2 = \int R(\nu) \, d\nu$, which results in a negligible dependence of $B$ on $N$ at the carrier densities of interest. The net power density that is deposited in the semiconductor is the difference between the power absorbed from the laser ($P_{abs}$) and that of the luminescence that escapes ($P_{le}$):

$$P_{net} = P_{abs} - P_{le} = [\alpha I + \Delta P] - [\eta_e B N^2 h\tilde{\nu}_f], \tag{1.14}$$

where the absorbed power density includes the resonant absorption ($\alpha I$) and a term $\Delta P$ that accounts for the undesirable effects such as free-carrier absorption and other parasitic absorptive processes. The second term is the escaped luminescence power density at a mean luminescence energy $h\tilde{\nu}_f$, defined as

$$h\tilde{\nu}_f = \frac{\int S(\nu) R(\nu) h\nu \, d\nu}{\int S(\nu) R(\nu) \, d\nu} . \tag{1.15}$$

Note that the escaped mean luminescence energy can deviate (i.e. redshift) from its internal value ($S = 1$) depending on the thickness or photon recycling conditions. With the aid of (1.9), we rewrite (1.14) as:

$$P_{net} = \eta_e B N^2 (h\nu - h\tilde{\nu}_f) + AN h\nu + CN^3 h\nu + \Delta P . \tag{1.16}$$

Equation 1.16 rigorously describes the laser cooling of a semiconductor in a compact and simple form. It accounts for the practical considerations of luminescence trapping by introducing an inhibited radiative recombination ($\eta_e B$) and a shifted mean photon energy $h\tilde{\nu}_f$ for the escaped luminescence. For high external efficiency systems where $S(\nu) = 1$, (1.16) approaches that described in the literature with $\eta_e = 1$ and $\tilde{\nu}_f = \nu_f$, where $\nu_f$ denotes the mean fluorescence energy produced internally in the semiconductor [44–46]. Equation 1.16 indicates that laser cooling occurs when $P_{net} < 0$, requiring a dominant contribution from the radiative recombination with $h\nu < h\tilde{\nu}_f$. The cooling efficiency $\eta_c$ is defined as the ratio of cooling power density $P_c (= -P_{net})$ to the absorbed laser power density ($P_{abs} = \alpha I + \Delta P$). With the aid of (1.11), this efficiency can be expressed as

$$\eta_c = -\frac{\eta_e BN^2 (h\nu - h\tilde{\nu}_f) + ANh\nu + CN^3 h\nu + \Delta P}{\eta_e BN^2 h\nu + ANh\nu + CN^3 h\nu + \Delta P}. \tag{1.17}$$

Ignoring the $\Delta P$ contributions for the moment, $\eta_c$ can be written more simply as:

$$\eta_c = \eta_{ext} \frac{\tilde{\nu}_f}{\nu} - 1, \tag{1.18}$$

where $\eta_{ext}$ describes the *external* quantum efficiency (or EQE):

$$\eta_{ext} = \frac{\eta_e BN^2}{AN + \eta_e BN^2 + CN^3} \approx (\eta_q)^{1/\eta_e}, \tag{1.19}$$

with $\eta_q = BN^2/(AN + BN^2 + CN^3)$ denoting the *internal* quantum efficiency [46,63], and defined more generally following (1.5). The approximate equality in (1.19) is valid only when $\eta_{ext}$ is close to unity (> 0.9). One simple consequence of (1.19) is that there is an optimum carrier density $N_{op} = (A/C)^{1/2}$ at which $\eta_{ext}$ reaches a maximum:

$$\eta_{ext}^{max} = 1 - \frac{2\sqrt{AC}}{\eta_e B}. \tag{1.20}$$

The inclusion of the background parasitic absorption ($\Delta P = \alpha_b I$) results in a more general form for the cooling efficiency:

$$\eta_c = \eta_{abs} \eta_{ext} \frac{\tilde{\nu}_f}{\nu} - 1, \tag{1.21}$$

where the absorption efficiency $\eta_{abs}$ is the fraction of all of the absorbed photons from the pump laser that are consumed by the resonant absorption in the cooling region:

$$\eta_{abs} = \frac{\alpha(\nu)}{\alpha(\nu) + \alpha_b}, \tag{1.22}$$

and $\alpha_b$ is assumed to be constant in the vicinity of the band-edge region.

If the pump laser suffers from parasitic absorption, so will the luminescence, since their frequencies are very close. We now examine the parasitic absorption problem and its effect on the cooling efficiency by revisiting (1.9). A small fraction $\varepsilon_f$ of the trapped luminescence is absorbed parasitically and the remaining part $(1 - \varepsilon_f)$ is recycled through interband absorption, thus contributing to carrier generation. Equation 1.9 is rewritten as:

$$\frac{dN}{dt} = \frac{\alpha I}{h\nu} - AN - BN^2 - CN^3 + (1 - \eta_e)(1 - \varepsilon_f) BN^2. \tag{1.23}$$

Note that $1 - \varepsilon_f = \bar{\alpha}_f/(\bar{\alpha}_f + \alpha_b) \approx 1 - \alpha_b/\bar{\alpha}_f$ where $\bar{\alpha}_f(\approx \alpha(\nu_f))$ is the interband absorption of the luminescence averaged over its spectrum. Following the same analysis leading to (1.21), we obtain the modified cooling efficiency:

$$\eta_c = \bar{\eta}_{ext} \eta_{abs} \frac{\tilde{\nu}_f}{\nu} - 1, \tag{1.24}$$

with a modified EQE ($\bar{\eta}_{ext}$) that is reduced from its ideal value in the high-purity ($\alpha_b \approx 0$) approximation to:

$$\bar{\eta}_{ext} = \eta_{ext} \frac{1}{1 + \eta_{ext}\varepsilon_f(1-\eta_e)/\eta_e} \approx \eta_{ext} - \eta_{ext}^2 \varepsilon_f (1-\eta_e)/\eta_e. \tag{1.25}$$

This expression is useful for setting an upper bound on the existing intrinsic background absorption of GaAs/InGaP heterostructures. This will be discussed in detail below. Parasitic luminescence absorption is not important in the analysis of photocarrier density and incident laser irradiance, so it is ignored for the moment.

The roots of (1.16) define the carrier density range within which net cooling can be observed provided that $\eta_e B(h\tilde{v}_f - hv) \geq 2hv\sqrt{AC}$. The equality defines the break-even condition: heating and cooling are in exact balance. At high quantum efficiency, radiative recombination dominates (i.e. $\eta_e B/C \gg N \gg A/\eta_e B$), allowing one to obtain the corresponding laser irradiance from (1.11) with the assumption of no band blocking. We can account for parasitic absorption of the pump by taking $\Delta P = \alpha_b I + \sigma_{fca} NI$, where $\alpha_b$ denotes an *effective* background parasitic absorption and $\sigma_{fca}$ is the free carrier absorption cross-section. In this case, net cooling can occur within an irradiance range of $I_1 < I < I_2$, where $I_{1,2} = (hv\eta_e B/\alpha(v))n_{1,2}^2$ and

$$N_{1,2} = \left(\frac{h\tilde{v}_f - hv}{hv} - \frac{\alpha_b}{\alpha(v)}\right) \frac{\eta_e B}{2C'} \left(1 \mp \sqrt{1 - \frac{A}{A_0}}\right). \tag{1.26}$$

Here $C' = C + \sigma_{fca}\eta_e B/\alpha(v)$, and

$$A_0 = \left(\frac{h\tilde{v}_f - hv}{hv} - \frac{\alpha_b}{\alpha(v)}\right)^2 \frac{(\eta_e B)^2}{4C'}, \tag{1.27}$$

is the break-even (maximum allowable) nonradiative decay rate for a given excitation energy $hv$. Free carrier absorption appears as an enhancement of the Auger process. The parameters $B$ and $C$ are fundamental properties of a semiconductor and have been calculated and measured extensively for various bulk and quantum-confined structures [59, 60, 63, 64]. The reported values for these coefficients, however, vary considerably. In bulk GaAs, for example, the published values are $2 \times 10^{-16} < B < 7 \times 10^{-16}$ m$^3$/s and $1 \times 10^{-42} < C < 7 \times 10^{-42}$ m$^6$/s [64]. These variations are primarily due to experimental uncertainties. We assume average values of $B = 4 \times 10^{-16}$ m$^3$/s and $C = 4 \times 10^{-42}$ m$^6$/s while ignoring the effects of background and free-carrier absorption. These assumptions allow us to gain insight into the feasibility and requirements for achieving net laser cooling. It should be noted that the theoretical values obtained for these parameters with different models vary within almost the same range as the experimental results. For the simple two-band model used here, $B \approx 5 \times 10^{-16}$ m$^3$/s [65].

Using (1.27), we plot in Figure 1.9 the break-even nonradiative lifetime $\tau_{nr}^0 = A_0^{-1}$ as a function of $\eta_e$ assuming $hv_f - hv = k_B T$, with $hv_f$ corresponding to $\lambda_f \approx 860$ nm

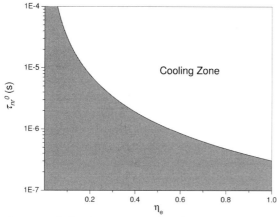

**Figure 1.9** The break-even nonradiative lifetime as a function of the luminescence extraction efficiency in bulk GaAs, calculated using typical values of radiative and Auger recombination at room temperature.

at room temperature. The orange area under the curve is the unwanted (heating) zone. Equation 1.27 also suggests that increasing the quantum efficiency $\eta_q$ by decreasing the incident photon energy (e.g. at $h\nu_f - h\nu > k_B T$) relaxes this requirement. Interband absorption drops rapidly as the excitation moves further into the Urbach tail and the background and free carrier absorption can no longer be ignored. Recently, it was found that the free-carrier absorption (FCA) at band-edge wavelengths is much smaller than previously expected [66,67]. For GaAs, $\sigma_{fca} \approx 10^{-24}$ m$^2$ [66,67], which requires that $\alpha(\nu) \geq 10^3$ m$^{-1}$ to ensure that free carrier losses are negligible (i.e. $C' \approx C$). This requirement is satisfied even at $\lambda = 890$ nm (corresponding to $h\nu_f - h\nu \approx 2k_B T$), where $\alpha(\nu) \approx 10^4$ m$^{-1}$. We conclude that FCA does not place a limitation on laser cooling.

We can categorize the possible sources and locations of the parasitic background absorption $\alpha_b$ into three regions: (a) active or core material, (b) cladding layers of the heterostructure, and (c) the substrate. It is also implicit that $\alpha_b$ in (1.27) is scaled such that, for cases (b) and (c), the actual background absorption coefficient $\alpha'_b = \alpha_b \times (d/L)$, where $d$ and $L$ are the thicknesses of the loss and active media (if different), respectively.

While situations (b) and (c) can be controlled experimentally by varying the barrier thickness or using high-purity substrate respectively, the parasitic absorption from the cooling layer itself presents the most difficult engineering obstacle. This limitation is revisited in the next section, where experiments on laser cooling with GaAs are analyzed.

It is also instructive to show an alternative and compact way of expressing the cooling condition. With laser excitation at $h\nu < h\bar{\nu}_f$, the cooling condition defined by EQE reduces to:

$$\eta_{ext} > \frac{\nu}{\bar{\nu}_f} + \frac{\alpha_b}{\alpha(\nu)}. \tag{1.28}$$

Including the parasitic absorption of luminescence allows us to replace $\eta_{\text{ext}}$ with $\tilde{\eta}_{\text{ext}}$, and (1.28) then gives a more general condition,

$$\tilde{\eta}_{\text{ext}} = \eta_{\text{ext}} - \frac{\alpha_b(1-\eta_e)}{\bar{\alpha}_f \eta_e} > \frac{\nu}{\nu_f} + \frac{\alpha_b}{\alpha(\nu)}. \quad (1.29)$$

The above inequality emphasizes the critical role of $\alpha_b$ in achieving net laser cooling. The quantity $\tilde{\eta}_{\text{ext}}$ can be measured accurately, so (1.28) defines the minimum value of EQE for a given background absorption, provided $\alpha(\nu)$ is known. The absorption $\alpha(\nu)$ drops sharply for energies considerably below the bandgap, which means that this inequality may never be satisfied for any wavelength if $\alpha_b$ is too large. To quantify this argument, we need to know the band-tail absorption accurately. The nature of the band-tail states and their dependence on the impurity type and concentration make the reported experimental values very sample specific. Most theoretical calculations are accurate only for above and near the bandgap wavelengths. It is best to approach the problem experimentally with absorption and luminescence data that allow accurate estimates of the required EQE using (1.9). Starting with the measured low-density photoluminescence (PL) spectrum on a high-quality sample, we obtain absorption spectra $\alpha(\nu)$ using the KMS relations of (1.13). The low-density approximation reduces the occupation factor to a simple Boltzmann factor, $\exp(-h\nu/k_B T)$, where we ignore possible band-filling (saturation) effects in the band tail. Using (1.9), the minimum $\eta_{\text{ext}}$ required can be estimated as a function of $h\nu$ for various values of $\alpha_b$, as depicted in Figure 1.10. Here we assume an extraction efficiency $\eta_e = 0.1$, which is typical of GaAs on a ZnS dome structure [15, 68].

Figure 1.10 indicates that the required EQE for cooling becomes more demanding as the temperature is lowered, which is essentially a consequence of the diminishing phonon population at low temperatures. This result mirrors the situation in

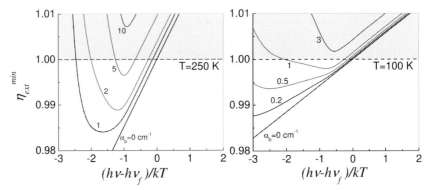

**Figure 1.10** The minimum EQE required to achieve laser cooling versus the normalized excitation photon energy for GaAs at $T = 250$ and 100 K, obtained from the inequality of (1.29). The absorption data $\alpha(\nu)$ were obtained by using the KMS relations on the PL spectra on a high-quality GaAs/InGaP double heterostructure. Note that for a certain background absorption $\alpha_b$, the requirement $\eta_{\text{ext}} > 1$ is unattainable (nonphysical) for any wavelength. This restriction becomes more prevalent at lower temperature.

the rare-earth-doped materials. Semiconductors, however, have the fortunate property that their EQE increases with decreasing temperature. The loss terms ($A$ and $C$ coefficients) decrease while the radiative rate ($B$ coefficient) increases inversely with lattice temperature. Using the accepted scaling for $C(T) \propto \exp(-\beta(300/T - 1))$ with $\beta \approx 2.4$ for GaAs [55,69], taking $B \propto T^{-3/2}$ [59,70], keeping $h\tilde{\nu}_f/h\nu - 1 \approx k_B T/E_g$, and ignoring parasitic losses and the small temperature dependence of the bandgap energy, we obtain for the break-even nonradiative decay rate

$$\frac{A_0(T)}{A_0(300)} \approx \left(\frac{300}{T}\right) \exp\left(\frac{\beta(300-T)}{T}\right). \tag{1.30}$$

At $T = 150$ K, for example, the break-even lifetime is ~ 40 times lower than it is at room temperature ($T = 300$ K). This is visualized by plotting $\eta_{\text{ext}}$ versus $T$, as shown in Figure 1.11 for two values of $\eta_e$. This range of values of $\eta_e$ corresponds to a GaAs structure bonded to a high refractive index dome of ZnS or ZnSe [15,68]. The solid red line indicates the break-even condition described by (1.2). This condition, together with the fact that $A$ (typically dominated by surface recombination) increases with temperature [70–72], makes the low-temperature observation of laser cooling more favorable even though the overall efficiency ($\approx k_B T/E_g$) decreases. The reduction in cooling efficiency is effectively due to the reduction of the electron–phonon absorption probability at lower temperatures. In particular, the population of LO phonons yields a corresponding reduction of the exciton linewidth $\Gamma$ [59]:

$$\Gamma(T) = \Gamma_0 + \sigma T + \gamma N_{\text{LO}}(T), \tag{1.31}$$

where $\Gamma_0$ is due to impurities and inhomogeneous broadening, $\sigma$ accounts for the contribution of acoustic phonons, $\gamma$ is the coefficient of LO-phonon scattering, and $N_{\text{LO}}(T)$ denotes the corresponding Bose–Einstein phonon distribution. For the exciton densities involved, we can ignore possible broadening due to exciton–exciton

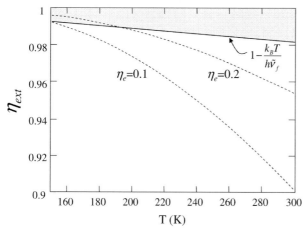

**Figure 1.11** The required external quantum efficiency (EQE) as a function of temperature for GaAs under typical parameters.

scattering [73]. As the lattice temperature approaches 10 K, the acoustic phonon contribution begins to dominate. At such low temperatures, however, the exciton–phonon scattering rate ($\approx \Gamma$) becomes comparable to the radiative recombination rate ($BN^2$) and consequently cold exciton recombination occurs before complete thermalization with the lattice. Similar processes that are related to premature hot exciton recombination have also hindered experimental observations of Bose–Einstein condensation in semiconductors. This problem is significantly alleviated by employing quantum-confined systems where $\sigma$ is enhanced by nearly three orders of magnitude. This relaxes wave-vector conservation along the confinement directions [74]. Enhanced cooling in quantum-confined systems may allow operation at temperatures < 10 K.

Another issue of concern is absorption saturation (band blocking) and many-body interactions. Band blocking may be a limiting factor for long-wavelength excitation where the low density of states gives rise to a stronger bleaching of the interband absorption. It is therefore necessary to have a good understanding of the absorption and emission spectra and their dynamic nonlinearities.

Theoretical models exist that deal with absorption spectra of semiconductor structures under various carrier densities and lattice temperatures. With different levels of complexity, there are theoretical calculations for 2D and 3D systems that deal with such many-body processes under dense e–h excitation [75–78]. Recently, a rigorous microscopic theory for absorption and luminescence in bulk semiconductors that includes the effects of electron–hole (e–h) plasma density as well as excitonic correlations has been introduced under the quasi-thermal equilibrium approximation [47, 79]. The reader is referred to the above sources for the details. Here, we use a simple model to estimate the effect of band blocking on achieving the carrier densities of (1.26) for GaAs. Using the electron–hole density of

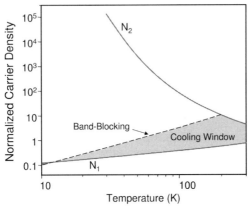

**Figure 1.12** The upper carrier density $N_2$, given by (1.26), is seen to be unattainable in GaAs due to band blocking as the temperature is lowered below 200 K. The *middle (dashed) line* represents the calculated density at which the band-tail absorption at $h\nu = h\nu_f - k_B T$ completely saturates (i.e. $\alpha(N) = 0$, (1.10)). This is a worst case scenario for which the nonradiative recombination rate is assumed to be constant with temperature.

states corresponding to a simple two-parabolic-band model, we calculate the carrier density at which the occupation factor $\{f_v - f_c\}$ vanishes at $h\nu = h\nu_f - k_B T$. The result is depicted in Figure 1.12, where this blocking density is contrasted with $N_1$ and $N_2$ of (1.26), evaluated using a constant $A$ coefficient and the temperature-dependent $B$ and $C$ coefficients used earlier. It is seen that band-blocking tends to reduce the cooling density window at $T < 200$ K and that cooling densities become unattainable at $T \approx 10$ K. The simple model overestimates band-blocking effects but qualitatively agrees with the more rigorous microscopic theory [47, 79]. Most recently, Khurgin presented a thorough analysis of the phonon-assisted band-tail states and the role of band blocking using a density matrix approach [52]. In that context, he pointed out the significance of the excitation wavelength in the presence of band blocking (saturation), which tends to diminish the cooling efficiency predominantly for $T \leq 100$ K [52].

Further details on the fundamental theory of laser cooling in semiconductors can be found in Chapters 6 and 7.

## 1.5
## Experimental Work on Optical Refrigeration in Semiconductors

The first thorough experimental effort was reported by the University of Colorado [46]. No net cooling was achieved, despite the realization of an impressive external quantum efficiency of 96% in GaAs. These experiments used a high-quality GaAs heterostructure that was optically contacted with a ZnSe dome structure for enhanced luminescence extraction. A report of local cooling in AlGaAs quantum wells by a European consortium [53] was later attributed to misinterpretation of spectra caused by Coulomb screening of the excitons [80]. Figure 1.8b displays anti-Stokes luminescence in a GaAs heterostructure, where excitation at $\lambda = 890$ nm produces broadband luminescence with a mean wavelength of $\lambda_f \approx 860$ nm. Each luminescent photon carries away about 40 meV more energy than an absorbed photon, so one might expect cooling. Why then have we not been able to observe laser cooling in this material or any semiconductor? To answer this question we have to revisit the cooling condition of (1.2), where strict requirements on EQE and background absorption are yet to be met. As discussed earlier, GaAs currently appears to be most promising due to the mature growth of this technology and its record quantum efficiency. There are, however, challenges that must be overcome to achieve the break-even cooling condition: (a) reduce the surface recombination rate $A$, (b) reduce the parasitic background absorption $\alpha_b$, and (c) enhance the luminescence extraction efficiency $\eta_e$. Issues (a) and (b) involve material preparation and both concern high-purity growth using advanced epitaxial methods. Condition (c), on the other hand, is a light management and device engineering challenge that deals with luminescence extraction from semiconductors with a high index of refraction. Total internal reflection causes most of the spontaneous emission to get trapped and reabsorbed. A similar problem limits the efficiency of light-emitting diodes (LEDs).

Various methods have been devised to remedy this problem for LEDs, but not all are applicable to laser cooling. For example, photon recycling in thin textured or in photonic bandgap structures can substantially enhance the luminescence extraction but at the price of redshifting the luminescence, as described by (1.15) [15]. Index-matched dome lenses have been exploited for LEDs and can be used for laser cooling as well [46], provided that the dome material does not introduce unacceptable levels of parasitic absorption. This requirement narrows the dome materials for GaAs to nearly index-matched ZnSe and ZnS due to the fact that they are currently available at a high purity grade [81]. GaP substrates or domes can provide a higher $\eta_e$ due to better index matching to GaAs, but currently available materials produce unacceptable levels of parasitic absorption [81]. Another method of improving $\eta_e$ makes use of nanosized vacuum gaps (nanogaps), and this will be briefly discussed later [6].

Highly controlled epitaxial growth techniques such as metal organic chemical vapor deposition (MOCVD) can produce very low surface recombination rates ($A < 10^4$ s$^{-1}$). This involves a double heterostructure of GaAs/GaInP, as shown in Figure 1.13, where the lattice-matched cladding layers provide surface passivation as well as carrier confinement. To deal with extraction efficiency, geometric coupling schemes such as nearly index-matched dome lenses (also shown in Figure 1.13) have been employed to enhance $\eta_e$ to 15–20%. The double heterostructures are lithographically patterned into ≈ 1 mm diameter disks that are lifted from their parent GaAs substrates and then bonded via van der Waals forces to a ZnS dome. The EQE of each sample is measured using the technique of fractional heating [46, 68] at various temperatures and laser pump powers.

The experimental setup is shown in Figure 1.14. The pump laser is a cw Ti:sapphire laser producing up to 4.5 W that is tunable in the wavelength range 750 – 900 nm. This laser is pumped by an 18 W laser at 532 nm (Verdi, Coherent Inc.). In the fractional heating experiment, the laser is tuned to around the mean luminescence wavelength while the temperature $\Delta T$ of the sample is measured.

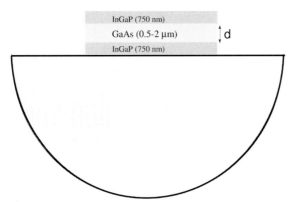

**Figure 1.13** The GaAs/GaInP double heterostructure is bonded to a nearly index-matched dome (ZnS or ZnSe) to enhance its luminescence extraction.

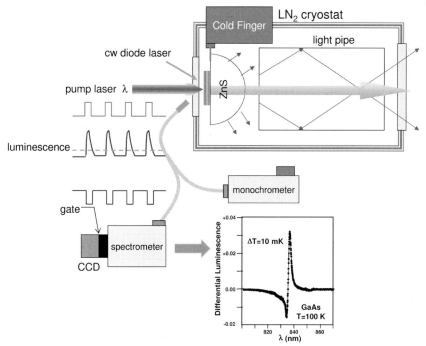

**Figure 1.14** Experimental setup for measuring the external quantum efficiency (EQE or $\eta_{ext}$) at various lattice temperatures in GaAs/GaInP double heterostructures. A tunable cw laser (Ti:sapphire, 4.5 W) excites a constant density of electron–hole pairs as the pump is tuned near the band edge. This is done by keeping the luminescence intensity (within a certain spectral band) constant using the monochrometer. After recombination, the temperature change of the sample is measured using differential luminescence thermometry (DLT), where the spectral shift of the low-density luminescence spectrum induced by a weak cw laser diode is monitored using the gated spectrometer. The *inset figure* shows a typical DLT signal for $\Delta T = 10$ mK in GaAs at $T = 100$ K.

According to the analysis described earlier, the temperature change in the sample is proportional to the net power deposited, which can be written as:

$$\Delta T(\lambda) = \kappa^{-1} P_{abs}(\lambda) \left(1 - \bar{\eta}_{ext}\eta_{abs}\frac{\lambda}{\lambda_f}\right), \tag{1.32}$$

where $\kappa$ is the total thermal conductance (W/K) of the system positioned in an optical cryostat. During the experiment, the absorbed power is kept constant: the pump laser is tuned and its power adjusted to keep the luminescence (or a fixed spectral portion of it) constant. This is shown in Figure 1.14, where a monochrometer is used to monitor the spectral portion of luminescence that does not overlap with the pump wavelength. The temperature change is monitored using a noncontact method called differential luminescence thermometry (DLT) that was developed for this purpose [68]. DLT exploits the temperature dependence of the luminescence resulting from the bandgap shift and broadening. High-temperature sensitivity is obtained by monitoring the differential spectrum before and after pump irradiation. To avoid complications caused by high carrier density [80], low-density luminescence spectra induced by a weak diode laser generate the DLT signals. This is achieved by modulating the pump laser with a mechanical chopper while synchronously gating a CCD camera on the spectrometer. This ensures that the DLT spectra are recorded when the pump laser is blocked and after the high-density luminescence has decayed. DLT signals are obtained by normalizing the signal and reference spectra before subtraction. The resulting differential signal is a peak–valley or valley–peak feature depending on the sign of $\Delta T$, which is calibrated in situ before an experiment. This method has exhibited a temperature resolution of better than 1 mK [68].

Returning to (1.32), it is evident that if $P_{abs}(\lambda)$ is kept constant, the measured $\Delta T$ versus $\lambda$ follows the wavelength dependence of the cooling (or heating) efficiency represented by the term in the parentheses. At short wavelengths ($\lambda < \lambda_f$) where $\alpha(\lambda)$ is large, $\eta_{abs}$ can be taken to be unity for moderate to high purity samples, and the fractional heating data follows $\Delta T(\lambda) \propto (1 - \bar{\eta}_{ext}\lambda/\lambda_f)$, which is a straight line. Currently available samples do not possess sufficient purity to make this term negative, i.e. net cooling. When the pump wavelength $\lambda$ is tuned close to $\lambda_f$, $\alpha_{abs}$ tends to degrade, thus preventing net cooling. Extrapolation of the short wavelength data can be used to obtain a "zero-crossing wavelength" $\lambda_c$, from which $\bar{\eta}_{ext} = \lambda_f/\lambda_c$ can be deduced. Recall that $\eta_{ext} > \bar{\eta}_{ext}$, and thus we obtain a lower limit on EQE. When parasitic absorption of luminescence can be ignored, $\eta_{ext} \approx \bar{\eta}_{ext}$. In the following analysis of the experimental data, we refer to this lower limit measured by fractional heating technique as the EQE. Figure 1.15a shows the measured EQE for various GaAs layer thicknesses at room temperature. The optimum GaAs thickness is found to be about 1 μm, determined by balancing excessive luminescence reabsorption for thicker layers with dominant surface recombination for thinner layers [55, 82]. The measured temperature dependence of EQE is depicted in Figure 1.15b, which is in qualitative agreement with the earlier analysis. Enhancement is observed as the temperature is lowered, reaching a record 99% at 100 K.

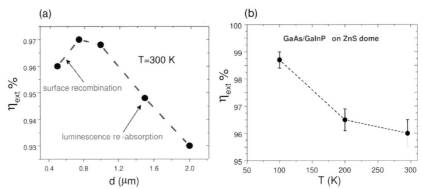

**Figure 1.15** External quantum efficiency (EQE) measured with (a) a bonded sample with various GaAs thicknesses, the GaInP layer thickness fixed at 0.5 μm, and $T = 300$ K, and (b) measured EQE vs. lattice temperature for a GaAs thickness of 1 μm.

The fractional heating data leading to 99% EQE is shown in Figure 1.16. We note the increase in temperature at longer wavelengths due to parasitic absorption. With the absorption of GaAs known, the data is fitted with a constant background of $\alpha_b \approx 10\,\text{cm}^{-1}$ assuming that it occurs entirely in the 1 μm thick GaAs layer. The experiment measures $\bar{\eta}_\text{ext}$, which means that the unmodified EQE ($\eta_\text{ext}$) can be even larger. The average absorption coefficient ($\bar{\alpha}_f \approx 4000\,\text{cm}^{-1}$) allows us to estimate an upper limit for $\alpha_b$ inside the double heterostructure. Even if $\eta_\text{ext} = 1$, attaining $\bar{\eta}_\text{ext} = 99\%$ requires that

$$\alpha_b < \frac{(1-\bar{\eta}_\text{ext})}{(1-\eta_e)/\eta_e}\bar{\alpha}_f \approx 4\,\text{cm}^{-1},$$

assuming that $\eta_e \approx 0.1$. The value of $\alpha_b = 10\,\text{cm}^{-1}$ that was used to fit the data in Figure 1.16 may include contributions from sources outside the double heterostructure, such as dome or cold-finger contacts.

Experiments show that the external quantum efficiency is sufficient to achieve net laser cooling, but the purity of the samples is not yet high enough. Efforts are underway to address this material problem. Researchers at the National Renewable Energy Laboratory are using highly controlled MOCVD growth, and scientists at the University of New Mexico are exploring aluminum-free GaAs heterostructure growth using phosphorus molecular beam epitaxy (MBE).

Methods of enhancing $\eta_\text{ext}$ by improving the luminescence extraction efficiency are also being explored. A novel method based on the frustrated total internal reflection across a vacuum "nano-gap" is being developed at the University of New Mexico [17, 83, 84]. In this scheme, the luminescence photons tunnel through the

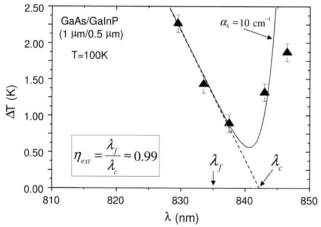

**Figure 1.16** Fractional heating of a 1 μm thick GaAs sample as a function of excitation wavelength for a fixed (optimum) electron–hole density at a starting temperature of 100 K. The extrapolation of short-wavelength data determines the zero-crossing wavelength, from which a record EQE of ~ 99% is measured. Excitation at longer wavelengths causes heating due to background parasitic absorption [6, 68].

(a)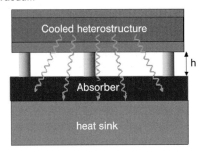

Vacuum

(b)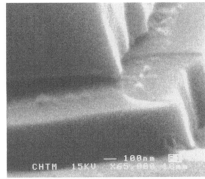

**Figure 1.17** (a) A vacuum "nanogap" structure where the heterostructure is situated (e.g. supported by posts) at a subwavelength distance from an absorber. Luminescence photons will escape into the absorber via frustrated total internal reflection (photon tunneling), while the gap provides a thermal barrier. (b) The SEM micrographs of a preliminary nanogap structure (with 50 nm spacing) fabricated using a multistep photolithographic technique [83].

gap into the absorber region. The vacuum nanogap maintains a thermal barrier between the heterostructure and the absorber/heat sink. The cooling heterostructure and the luminescence absorber are thus optically contacted but thermally insulated. Calculations show that a gap spacing of < 25 nm has a higher extraction efficiency than the dome structure (GaAs on ZnS or ZnSe), and the preliminary fabrication of such structures has shown promising results. Using a multistep photolithographic process, a Si-based nanogap with 50 nm spacing supported by posts has been monolithically fabricated, as depicted in Figure 1.17 [83]. Recently, we fabricated GaAs nanogap structures that will be integrated with a high-quality GaAs heterostructure to investigate their performance in cooling experiments.

## 1.6
### Future Outlook

Optical refrigeration has advanced from basic principles to a promising technology. The cooling of rare-earth-based materials is approaching cryogenic operation. In semiconductors, much progress has been made in achieving high external quantum efficiency. With advanced heterostructure growth and novel device fabrication currently underway, cooling will soon be attainable. In the coming years, optical refrigeration will be useful in applications such as satellite instrumentation and small sensors, where compactness, ruggedness and a lack of vibrations are important. Optical refrigeration is well suited to spaceborne applications since it does not involve any moving parts and can be designed for long operational lifetimes. Additionally, the cooling element is not electrically powered, so it will not interfere with the electronics being cooled. One can envision optical refrigerators being directly integrated with infrared sensors for thermal imaging of the Earth and of astrophysical objects. In terrestrial applications, their small size and high reliability make optical refrigerators attractive for use with high-temperature superconductor sensors and electronics. Optical refrigeration could enable compact SQUID magnetometers for geophysical and biomedical sensing, and may be a critical component in the production of commercial electronics incorporating fast and efficient superconducting components.

## Appendix I: Photon Waste Recycling and the Carnot Limit

Many applications will be possible if the basic efficiency of optical refrigerators $\sim k_B T/h\nu$ can be improved. This efficiency limit assumes that all fluorescence photons are absorbed by a heat sink and thus wasted. This is depicted diagrammatically in Figure 1.18a. One expects that overall efficiency will improve if fluorescence photons are recycled with photovoltaic (PV) elements in order to convert them into electricity [85]. This recovered energy can be used to drive laser diodes (LD) at the pump photon energy $h\nu$. This process is shown in Figure 1.18b.

Deriving the new cooling efficiency is then a straightforward process. For the same cooling power ($P_c$) as in the open loop system, the required laser power in the photon-waste recycled system is lowered by the amount generated by recycling, $\eta_{PV}\eta_L P_f$, where $P_f$ is the luminescence power and $\eta_{PV}$ and $\eta_L$ are the power efficiencies of the photovoltaic and the laser diode, respectively. The enhanced cooling efficiency $\tilde{\eta}_c$ is then obtained as

$$\tilde{\eta}_c = \frac{\eta_c}{1 - \eta_{PV}\eta_L(1-\eta_c)}, \qquad (1.33)$$

where the "old" or open loop cooling efficiency $\eta_c = 1 - P_f/P_L$, as before. It is useful to investigate what limits this new efficiency. Assuming that $\eta_c$ is given by its quantum limit of $h\nu_f/h\nu - 1$, we take $h\nu = h\nu_f - mk_B T_c$, where $m$ is on the order of unity and $T_c$ is the temperature of the solid-state coolant. This leads to

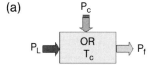

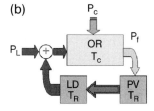

**Figure 1.18** Diagrams of (a) an open loop optical refrigerator (OR) and (b) a closed loop system in which the luminescence (photon waste) recycling is performed using photovoltaic (PV) elements followed by a laser diode (LD), both at a temperature $T_R$.

$\eta_c = mk_B T_c/h\nu_f \approx mk_B T_c/E_g$, where $E_g$ is the energy gap of the transition. The obvious choice for the photovoltaic and laser diode system would be a semiconductor with an energy gap that is very close to $E_g$. Assuming that both are at a reservoir temperature $T_R$, it is not unreasonable to take their conversion efficiencies to be $\eta_{PV} = 1 - pk_B T_R/E_g$ and $\eta_L = 1 - qk_B T_R/E_g$, respectively. Current devices reach efficiencies that correspond to $p \approx q \approx$ 30–50, so one expects they will be limited by $p \approx q \approx 1$. For such a limit, (1.33) can be written as:

$$\bar{\eta}_c = \frac{T_C}{T_C + \left(\frac{p+q}{m}\right) T_R}. \tag{1.34}$$

We note that thermodynamic analysis requires that $p + q \geq m \approx 1$. With $p + q = m \approx 1$, one obtains the Carnot limit of $T_C/(T_C + T_R)$. Expected technological advances in photovoltaic and laser diode devices will enable efficiencies approaching that corresponding to $p \approx q \approx 1$, and so the ultimate efficiency of the optical refrigerator will be given by its Carnot limit. Further discussions on the thermodynamics of solid-state laser cooling can be found in Chapter 8 by Mungan.

## References

1 CHU, S., COHEN-TANNOUDJI, C. AND PHILIPS, W.D. (1997) Nobel Prize in Physics for the development of methods to cool and trap atoms with laser light.

2 CORNELL, E.A., KETTERLE, W. AND WEIMAN, C.E. (2001) Nobel Prize in Physics for the achievement of Bose–Einstein condensation in dilute gases of alkali atoms, and for early fundamental studies of the properties of the condensates.

3 PRINGSHEIM, P. (1929) Zwei Bemerkungen über den Unterchied von Lumineszenz- und Temperaturstrahlung, Z. Phys., 57, 739.

4 EDWARDS, B.C., BUCHWALD, M.I. AND EPSTEIN, R.I. (1998) Development of the Los-Alamos Solid-State Optical Refrigerator, Rev. Sci. Instrum., 69, 2050–2055.

5 EDWARDS, B.C., ANDERSON, J.E., EPSTEIN, R.I., MILLS, G.L. AND MORD, A.J. (1999) Demonstration of a solid-state optical cooler: An approach to cryogenic refrigeration, J. Appl. Phys., 86, 6489–6493.

6 SHEIK-BAHAE, M. AND EPSTEIN, R.I. (2007) Optical refrigeration, Nature Photonics, 1, 693–699.

7 MILLS, G. AND MORD, A. (2005) Performance modeling of optical refrigerators, Cryogenics, 46, 176–182.

8 BOWMAN, S.R. (1999) Lasers without internal heat generation, IEEE J. Quant. Elect., 35, 115–122.

9 BOWMAN, S.R. AND MUNGAN, C.E. (2000) New materials for optical cooling, Applied Physics B, 71, 807–811.

10 LANDAU, L. (1946) On the thermodynamics of photoluminescence, J. Phys. (Moscow), 10, 503.

11 KASTLER, A. (1950) Quelques suggestions concernant la production optique et la détection optique d'une inégalité de population des niveaux de quantification spatiale des atomes application a l'expérience de Stern et Gerlach et a la résonance magnétique, J. Phys. Radium, 11, 255–265.

12 Kushida, T. and Geusic, J.E. (1968) Optical refrigeration in Nd-doped yttrium aluminum garnet, Phys. Rev. Lett., 21, 1172–1175.

13 YATSIV, S. (1961) Anti-Stokes fluorescence as a cooling process, in Advances in Quantum Electronics, (ed J.R. Singer), New York, Columbia Univ., 200–213.

14 Epstein, R.I., Buchwald, M.I., Edwards, B.C., Gosnell, T.R. and Mungan, C.E. (1995) Observation of laser-induced fluorescent cooling of a solid, *Nature*, **377**, 500–503.
15 Sheik-Bahae, M. and Epstein, R.I. (2004) Can laser light cool Semiconductors?, *Phys. Rev. Lett.*, **92**, 247403.
16 Asbeck, P. (1977) Self-absorption effects on radiative lifetime in GaAs–GaAlAs Double Heterostructures, *J. Appl. Phys.*, **48**, 820-822.
17 Sheik-Bahae, M., Imangholi, B., Hasselbeck, M.P., Epstein, R.I. and Kurtz, S. (2006) Advances in laser cooling of semiconductors, in *Proceedings of SPIE: Physics and Simulation of Optoelectronic Devices XIV*, vol. 6115, (eds M. Osinski, F. Henneberger and Y. Arakawa), Bellingham, SPIE, pp. 611518.
18 Emin, D. (2007) Laser cooling via excitation of localized electrons, *Phys. Rev. B*, **76**, 024301.
19 Mungan, C.E., Buchwald, M.I., Edwards, B.C., Epstein, R.I. and Gosnell, T.R. (1997) Laser cooling of a solid by 16 K starting from room-temperature, *Phys. Rev. Lett.*, **78**, 1030–1033.
20 Luo, X., Eisaman, M.D. and Gosnell, T.R. (1998) Laser cooling of a solid by 21 K starting from room temperature, *Opt. Lett.*, **23**, 639–641.
21 Gosnell, T.R. (1999) Laser cooling of a solid by 65 K starting from room temperature, *Opt. Lett.*, **24**, 1041–1043.
22 Thiede, J., Distel, J., Greenfield, S.R. and Epstein, R.I. (2005) Cooling to 208 K by optical refrigeration, *Appl. Phys. Lett.*, **86**, 154107.
23 Rayner, A., Friese, M.E.J., Truscott, A.G., Heckenberg, N.R. and Rubinsztein-Dunlop, H. (2001) Laser cooling of a solid from ambient temperature, *J. Mod. Opt.*, **48**, 103–114.
24 Heeg, B., Stone, M.D., Khizhnyak, A., Rumbles, G., Mills, G. and DeBarber, P.A. (2004) Experimental demonstration of intracavity solid-state laser cooling of $Yb^{3+}$: $ZrF_4$–$BaF_2$–$LaF_3$–$AlF_3$–NaF glass, *Phys. Rev. A*, **70**, 021401.
25 Fernandez, J.R., Mendioroz, A., Balda, R., Voda, M., Al-Saleh, M., Garcia-Adeva, A.J., Adam, J.-L. and Lucas, J. (2002) Origin of laser-induced internal cooling of $Yb^{3+}$-doped systems, Rare-Earth-Doped Materials and Devices VI, vol. 4645, S. Jiang and R.W. Keys, Eds., Bellingham, SPIE, pp. 135–147.
26 Fernandez, J., Mendioroz, A., Garcia, A.J., Balda, R. and Adam, J.L. (2000) Anti-Stokes laser-induced internal cooling of $Yb^{3+}$-doped glasses, *Phys. Rev. B*, **62**, 3213–3217.
27 Epstein, R.I., Brown, J.J., Edwards, B.C. and Gibbs, A. (2001) Measurements of optical refrigeration in ytterbium-doped crystals, *J. Appl. Phys.*, **90**, 4815–4819.
28 Mendioroz, A., Fernandez, J., Voda, M., Al-Saleh, M., Balda, R. and Garcia-Adeva, A. (2002) Anti-stokes laser cooling in $Yb^{3+}$-doped $KPb_2Cl_5$ crystal, *Opt. Lett.*, **27**, 1525–1527.
29 Bigotta, S., Parisi, D., Bonelli, L., Toncelli, A., Di Lieto, A. and Tonelli, M. (2006) Laser cooling of $Yb^{3+}$-doped $BaY_2F_8$ single crystal, *Opt. Mater.*, **28**, 1321–1324.
30 Bigotta, S., Parisi, D., Bonelli, L., Toncelli, A., Tonelli, M. and Di Lieto, A. (2006) Spectroscopic and laser cooling results on $Yb^{3+}$-doped $BaY_2F_8$ single crystal, *J. Appl. Phys.*, **100**, 013109.
31 Patterson, W., Bigotta, S., Sheik-Bahae, M., Parisi, D., Tonelli, M. and Epstein, R. (2008) Anti-Stokes luminescence cooling of $Tm^{3+}$-doped $BaY_2F_8$, *Opt. Express*, **16**, 1704–1710.
32 Bigotta, S., Lieto, A.D., Parisi, D., Toncelli, A. and Tonelli, M. (2007) *Single fluoride crystals as materials for laser cooling applications*, presented at Laser Cooling of Solids, San Jose, CA, USA, 24–25 Jan.
33 Seletskiy, D., Hasselbeck, M.P., Sheik-Bahae, M., Epstein, R.I. Bigotta, S. and Tonelli, M. (2008) Cooling of Yb:YLF using cavity enhanced resonant absorption, presented at Laser Refrigeration of Solids, San Jose, CA, USA, 23–24 Jan.
34 Hoyt, C.W., Sheik-Bahae, M., Epstein, R.I., Edwards, B.C. and Anderson, J.E. (2000) Observation of anti-Stokes fluorescence cooling in thulium-doped glass, *Phys. Rev. Lett.*, **85**, 3600–3603.
35 Hoyt, C., Hasselbeck, M., Sheik-Bahae, M., Epstein, R., Greenfield, S., Thiede, J., Distel, J. and Valencia, J. (2003) Ad-

vances in laser cooling of thulium-doped glass, *J. Opt. Soc. Am. B*, **20**, 1066–1074.

36 FERNANDEZ, J., GARCIA-ADEVA, A.J. AND BALDA, R. (**2006**) Anti-Stokes laser cooling in bulk erbium-doped materials, *Phys. Rev. Lett.*, **97**, 033001.

37 GARCIA-ADEVA, A.J., BALDA, R. AND FERNANDEZ, J. (**2007**) Anti-Stokes laser cooling in erbium-doped low-phonon materials, in *Laser Cooling of Solids*, vol. 6461, (eds R.I. Epstein and M. Sheik-Bahae), Bellingham, SPIE, pp. 646102-1.

38 LAMOUCHE, G., LAVALLARD, P., SURIS, R. AND GROUSSON, R. (**1998**) Low temperature laser cooling with a rare-earth doped glass, *J. Appl. Phys.*, **84**, 509–516.

39 HEHLEN, M.P., EPSTEIN, R.I. AND INOUE, H. (**2007**) Model of laser cooling in the $Yb^{3+}$-doped fluorozirconate glass ZBLAN, *Phys. Rev. B*, **75**, 144302.

40 HOYT, C.W. (**2003**) Laser Cooling in Thulium-Doped Solids, Ph.D. Thesis, University of New Mexico, 138 pp.

41 SELETSKIY, D., HASSELBECK, M.P., SHEIK-BAHAE, M. AND EPSTEIN, R.I. (**2007**) Laser cooling using cavity enhanced pump absorption, in *Proceeding of SPIE: Laser Cooling of Solids*, vol. 6461, (eds R.I. Epstein and M. Sheik-Bahae), Bellingham, SPIE, 646104.

42 RUAN, X.L. AND KAVIANY, M. (**2006**) Enhanced laser cooling of rare-earth-ion-doped nanocrystalline powders, *Phys. Rev. B*, **73**, 155422.

43 CLARK, J.L. AND RUMBLES, G. (**1996**) Laser cooling in the condensed-phase by frequency up-conversion, *Phys. Rev. Lett.*, **76**, 2037–2040.

44 ORAEVSKY, A.N. (**1996**) Cooling of semiconductors by laser radiation, *J. Russ. Laser Res.*, **17**, 471–479.

45 RIVLIN, L.A. AND ZADERNOVSKY, A.A. (**1997**) Laser cooling of semiconductors, *Opt. Commun.*, **139**, 219–222.

46 GAUCK, H., GFROERER, T.H., RENN, M.J., CORNELL, E.A. AND BERTNESS, K.A. (**1997**) External radiative quantum efficiency of 96% from a GaAs/GaInP heterostructure, *Appl. Phys. A*, **64**, 143–147.

47 RUPPER, G., KWONG, N.H. AND BINDER, R. (**2006**) Large excitonic enhancement of optical refrigeration in semiconductors, *Phys. Rev. Lett.*, **97**, 117401.

48 HUANG, D.H., APOSTOLOVA, T., ALSING, P.M. AND CARDIMONA, D.A. (**2005**) Spatially selective laser cooling of carriers in semiconductor quantum wells, *Phys. Rev. B*, **72**, 195308.

49 LI, J.Z. (**2007**) Laser cooling of semiconductor quantum wells: Theoretical framework and strategy for deep optical refrigeration by luminescence upconversion, *Phys. Rev. B*, **75**, 155315.

50 RUPPER, G., KWONG, N.H. AND BINDER, R. (**2007**) Optical refrigeration of GaAs: Theoretical study, *Phys. Rev. B*, **76**, 245203.

51 HUANG, D. AND ALSING, P.M. (**2008**) Many-body effects on optical carrier cooling in intrinsic semiconductors at low lattice temperatures, *Phys. Rev. B*, **78**, 035206.

52 KHURGIN, J.B. (**2008**) Role of bandtail states in laser cooling of semiconductors, *Phys. Rev. B*, **77**, 235206.

53 FINKEISSEN, E., POTEMSKI, M., WYDER, P., VINA, L. AND WEIMANN, G. (**1999**) Cooling of a semiconductor by luminescence up-conversion, *Appl. Phys. Lett.*, **75**, 1258–1260.

54 GFROERER, T.H., CORNELL, E.A. AND WANLASS, M.W. (**1998**) Efficient directional spontaneous emission from an InGaAs/InP heterostructure with an integral parabolic reflector, *J. Appl. Phys.*, **84**, 5360–5362.

55 IMANGHOLI, B., HASSELBECK, M.P., SHEIK-BAHAE, M., EPSTEIN, R.I. AND KURTZ, S. (**2005**) Effects of epitaxial lift-off on interface recombination and laser cooling in GaInP/GaAs heterostructures, *Appl. Phys. Lett.*, **86**, 81104.

56 ESHLAGHI, S., WORTHOFF, W., WIECK, A.D. AND SUTER, D. (**2008**) Luminescence upconversion in GaAs quantum wells, *Phys. Rev. B*, **77**, 245317.

57 KHURGIN, J.B. (**2007**) Surface plasmon-assisted laser cooling of solids, *Phys. Rev. Lett.*, **98**, 177401.

58 KHURGIN, J.B. (**2006**) Band gap engineering for laser cooling of semiconductors, *J. Appl. Phys.*, **100**, 113116.

59 BASU, P.K. (**1997**) Theory of Optical Processes in Semiconductors: Bulk and Microstructures, New York: Oxford Univ. Press.

60 PANKOVE, J.I. (**1971**) Optical Processes in Semiconductors, Englewood Cliffs, Prentice-Hall.
61 BORN, M. AND WOLF, E. (**1999**) Principles of Optics: Electromagnetic Theory of Propagation, Interference and Diffraction of Light, 7th expanded ed., Cambridge, Cambridge Univ. Press.
62 VAN ROOSBROECK, W. AND SHOCKLEY, W. (**1954**) *Phys. Rev.*, **94**, 1558.
63 CHUANG, S.L. (**1995**) Physics of Optoelectronic Devices, New York, John Wiley & Sons, Inc.
64 PIPREK, J. (**2003**) Semiconductor Optoelectronic Devices: Introduction to Physics and Simulation, Amsterdam, Academic.
65 YABLONOVITCH, E., GMITTER, T.J. AND BHAT, R. (**1988**) Inhibited and enhanced spontaneous emission from optically thin AlGaAs/GaAs double heterostructures, *Phys. Rev. Lett.*, **61**, 2546–2549.
66 HAUG, A. (**1992**) Free-carrier absorption in semiconductor-lasers, *Semicond. Sci. Tech.*, **7**, 373–378.
67 YI, H., DIAZ, J., LANE, B. AND RAZEGHI, M. (**1996**) Optical losses of Al-free lasers for $\lambda$ = 0.808 and 0.98 µm, *Appl. Phys. Lett.*, **69**, 2983–2985.
68 IMANGHOLI, B. (**2006**) Investigation of laser cooling in semiconductors, Ph.D. Thesis, University of New Mexico, 210 pp.
69 AGRAWAL, G.P. AND DUTTA, N.K. (**1986**) Long-wavelength semiconductor lasers, New York, Van Nostrand Reinhold.
70 THOOFT, G.W. AND VANOPDORP, C. (**1983**) Temperature-dependence of interface recombination and radiative recombination in (AL,GA)AS heterostructures, *Appl. Phys. Lett.*, **42**, 813–815.
71 AHRENKIEL, R.K., OLSON, J.M., DUNLAVY, D.J., KEYES, B.M. AND KIBBLER, A.E. (**1990**) Recombination velocity of the $Ga_{0.5}In_{0.5}P$/GaAs interface, *J. Vac. Sci. Technol. A*, **8**, 3002–3005.
72 OLSON, J.M., AHRENKIEL, R.K., DUNLAVY, D.J., KEYES, B. AND KIBBLER, A.E. (**1989**) Ultralow recombination velocity at $Ga_{0.5}In_{0.5}P$/GaAs heterointerfaces, *Appl. Phys. Lett.*, **55**, 1208–1210.
73 SCHAEFER, A.C. AND STEEL, D.G. (**1997**) Nonlinear optical response of the GaAs exciton polariton, *Phys. Rev. Lett.*, **79**, 4870–4873.
74 BUTOV, L.V., LAI, C.W., IVANOV, A.L., GOSSARD, A.C. AND CHEMLA, D.S. (**2002**) Towards Bose–Einstein condensation of excitons in potential traps, *Nature*, **417**, 47–52.
75 BANYAI, L. AND KOCH, S.W. (**1986**) A simple theory for the effects of plasma screening on the optical-spectra of highly excited semiconductors, *Z. Phys. B*, **63**, 283–291.
76 HAUG, H. AND KOCH, S.W. (**1994**) Quantum Theory of the Optical and Electronic Properties of Semiconductors, 3rd edn, Singapore: World Scientific.
77 LOWENAU, J.P., REICH, F.M. AND GORNIK, E. (**1995**) Many-body theory of room-temperature optical nonlinearities in bulk semiconductors, *Phys. Rev. B*, **51**, 4159–4165.
78 MEIER, T. AND KOCH, S.W. (**2001**) Coulomb correlation signatures in the excitonic optical nonlinearities of semiconductors, *Semicond. Semimet.*, **67**, 231–313.
79 RUPPER, G., KWONG, N.H. AND BINDER, R. (**2007**) Optical refrigeration of GaAs: Theoretical study, *Phys. Rev. B*, **76**, 245203.
80 HASSELBECK, M.P., SHEIK-BAHAE, M. AND EPSTEIN, R.I. (**2007**) Effect of high carrier density on luminescence thermometry in semiconductors, in *Proceedings of SPIE: Laser Cooling of Solids*, vol. 6461, (eds R.I. Epstein and M. Sheik-Bahae), Bellingham, SPIE, 646107.
81 IMANGHOLI, B., HASSELBECK, M.P. AND SHEIK-BAHAE, M. (**2003**) Absorption spectra of wide-gap semiconductors in their transparency region, *Opt. Commun.*, **227**, 337–341.
82 CATCHPOLE, K.R., LIN, K.L., CAMPBELL, P., GREEN, M.A., BETT, A.W. AND DIMROTH, F. (**2004**) High external quantum efficiency of planar semiconductor structures, *Semicond. Sci. Tech.*, **19**, 1232–1235.
83 MARTIN, R.P., VELTEN, J., STINTZ, A., MALLOY, K.J., EPSTEIN, R.I., SHEIK-BAHAE, M., HASSELBECK, M.P., IMANGHOLI, B., BOYD, S.T.P. AND BAUER, T.M. (**2007**) Nanogap experiments for laser cooling: a progress report, in *Proceedings of SPIE: Laser Cooling of Solids*, vol. 6461, (eds R.I. Epstein

and M. Sheik-Bahae), Bellingham, SPIE, 64610H.

**84** EPSTEIN, R.I., EDWARDS, B.C. AND SHEIK-BAHAE, M. **(2002)** Semiconductor-Based Optical Refrigerator, US Patent 6 378 321.

**85** EDWARDS, B.C., BUCHWALD, M.I. AND EPSTEIN, R.I. **(2000)** Optical refrigerator using reflectivity tuned dielectric mirrors, US Patent 6 041 610.

# 2
# Design and Fabrication of Rare-Earth-Doped Laser Cooling Materials
*Markus P. Hehlen*

## 2.1
### History of Laser Cooling Materials

In 1929, Peter Pringsheim suggested for the first time that it should be possible to cool a fluorescent gas through its interaction with radiation [1]. He envisioned a system that absorbs light at one wavelength and then re-emits it at a slightly shorter wavelength, in a process called anti-Stokes fluorescence. The energy difference between the excitation light and the anti-Stokes fluorescence would be supplied by the thermal reservoir of the gas and, as a result, induce a net cooling of the object. Pringsheim also argued that this process would not violate the second law of thermodynamics, a fact that was proven formally by Landau in 1946 [2]. A few years later, Kastler proposed that gas-phase atoms as well as rare-earth ions in transparent solids could be fluorescent coolers because of their high quantum efficiencies and narrow spectral lines [3]. The difference between the excitation and anti-Stokes fluorescence energies tends to be on the order of only a few $kT$ and thus requires a spectrally narrow excitation source. It was not until the emergence of the laser in the 1960s that excitation light with such a narrow spectral power distribution could be supplied to realize effective laser cooling. By the mid-1980s, gas-phase atoms and ions had been cooled to milli-Kelvin temperatures by means of Doppler cooling [4], an effect that has since grown into a sizeable research area. In contrast, progress towards cooling a solid with light was slower. In 1968, Kushida and Geusic reported the first experimental attempt to cool a solid with laser light [5]. They excited a $Nd^{3+}$-doped $Y_3Al_5O_{12}$ (YAG) crystal at 1064 nm to produce anti-Stokes fluorescence at 946 nm, and they ascribed the less-than-expected heating to optical refrigeration. While net cooling was not achieved, Kushida and Geusic recognized the critical role of nonradiative relaxation – both of the excited active ion itself and of excited impurity species – in degrading the efficiency of laser cooling.

A breakthrough occurred in 1995 when Epstein *et al.* experimentally demonstrated the net cooling of a solid by light using $Yb^{3+}$-doped fluorozirconate glass (ZBLAN:$Yb^{3+}$) [6]. This material was chosen to minimize nonradiative relaxation processes that degrade the laser cooling efficiency. In particular, the nonradiative

*Optical Refrigeration. Science and Applications of Laser Cooling of Solids.*
Edited by Richard Epstein and Mansoor Sheik-Bahae
Copyright © 2009 WILEY-VCH Verlag GmbH & Co. KGaA, Weinheim
ISBN: 978-3-527-40876-4

relaxation of the Yb$^{3+}$ excited state ($^2F_{5/2}$) to the Yb$^{3+}$ ground state ($^2F_{7/2}$) via interactions with the optical phonons of the ZBLAN glass host has a probability of essentially zero because the $^2F_{5/2}$–$^2F_{7/2}$ energy difference of ~ 9770 cm$^{-1}$ is large compared to the highest-energy optical phonon in ZBLAN (~ 580 cm$^{-1}$) [7] (see Section 2.2.2). Epstein et al. measured net laser cooling efficiencies of up to 2% when they excited the material with a laser at around 1030 nm and room temperature. Realization of solid-state laser cooling in the mid-1990s coincided with an intense effort in academia and industry to develop and commercialize ultra-low-loss fluoride glass fibers for long-haul fiber-optic telecommunications. The focus was on heavy metal fluoride glasses (HMFG), and ZBLAN (ZrF$_4$-BaF$_2$-LaF$_3$-AlF$_3$-NaF) glass in particular. Such glasses have a predicted minimum light attenuation of only ~ 0.01 dB/km (at 2.5 µm) [8], which is ten times lower than the theoretical attenuation minimum of silica, ~ 0.1 dB/km (at 1.5 µm). The field was fueled by the prospect of realizing transoceanic fluoride fiber-optic links without the need for signal amplification, which is not possible with silica fibers. ZBLAN turned out to be a substantially more difficult material to work with than silica, and various intrinsic material properties prevented it from ever approaching its theoretical low attenuation loss in long fibers. The realization of low-loss (~ 0.25 dB/km) silica fibers, together with the emergence of high-performance erbium-doped fiber amplifiers (EDFAs) operating around 1.55 µm, paved the way for the global deployment of silica-based fiber-optic networks during the "boom" years (1999–2001) in telecommunications. Fluoride glasses were being pushed into niche applications for which relatively short lengths of fiber were adequate. Such applications include fiber-optic sensors, fiber-optic amplifiers, fiber lasers, and fiber delivery of light in laser surgery. Nevertheless, research into laser cooling benefited greatly from this effort. High-quality ZBLAN:Yb$^{3+}$ samples from fiber preforms were available commercially at the time, and ZBLAN:Yb$^{3+}$ became the workhorse for laser cooling. Over a dozen experimental studies of laser cooling in this material had been published by 2005 (see Table 2.1). This effort resulted in the laser cooling of a ZBLAN:Yb$^{3+}$ bulk sample to 208 K from room temperature [9], which is the record low temperature for the optical refrigeration of a solid to date.

The year 2000 marked important new directions in the field of solid-state optical refrigeration: laser cooling was observed for the first time (1) in a Tm$^{3+}$ system (ZBLAN:Tm$^{3+}$), (2) in new fluoride (BIG:Yb$^{3+}$) and fluorochloride (CNBZn:Yb$^{3+}$) glasses, and (3) in crystals (KGd(WO$_4$)$_2$:Yb$^{3+}$ and KY(WO$_4$)$_2$:Yb$^{3+}$). The search for new systems continues to this day and is driven primarily by (1) the difficulty in obtaining commercial glasses and crystals of sufficient purity and quality for laser cooling, (2) the need for materials that are less laborious to fabricate than ZBLAN at exceedingly high purities, and (3) the prospect of realizing higher laser cooling efficiencies in Tm$^{3+}$ compared to Yb$^{3+}$ materials. Steady progress in recent years has resulted in laser cooling by 24 K in ZBLAN:Tm$^{3+}$ glass and by 69 K in a YLiF$_4$:Yb$^{3+}$ crystal from room temperature. So far, solid-state laser cooling has been observed in 15 different glasses and crystals and with the anti-Stokes fluorescent ions Yb$^{3+}$, Tm$^{3+}$, and Er$^{3+}$ (see Table 2.1).

**Table 2.1** Compilation of rare-earth-doped glasses and crystals in which net optical refrigeration has been observed to date. The initial sample temperature, $T$, and the reported maximum temperature drop, $\Delta T$, are given along with the method by which $\Delta T$ was determined. These temperatures are for reference only and are not directly comparable across all thermometry methods and experimental configurations reported in this table. The thermometry methods are thermocouple (TC), infrared camera (IR), photothermal deflection (PTD), and differential luminescence thermometry (DLT).

| Material class | Active ion | Material | $T$ [K] | $\Delta T$ [K] | Method | Reference |
|---|---|---|---|---|---|---|
| **Glasses** | | | | | | |
| Fluorides | $Yb^{3+}$ | ZBLAN ($ZrF_4$-$BaF_2$-$LaF_3$-$AlF_3$-NaF) | RT | 92 | TC | [9] |
| | | | 301 | 65 | DLT | [10] |
| | | | RT | 48 | TC | [11] |
| | | | 298 | 21 | DLT | [12] |
| | | | 298 | 16 | PTD | [13] |
| | | | RT | 13 | DLT | [14] |
| | | | RT | 7.9 | TC | [15] |
| | | | RT | 6 | DLT | [16] |
| | | | RT | 4 | PTD | [17] |
| | | | RT | 3.7 | TC | [18] |
| | | | RT | 3 | DLT | [19] |
| | | | RT | 0.33 | PTD, IR | [20] |
| | | | RT | 0.3 | PTD, IR | [6] |
| | | | 300–100 | – | PTD | [21] |
| | $Yb^{3+}$ | BIG ($BaF_2$-$InF_3$-$GaF_3$-$ZnF_2$-$LuF_3$-$GdF_3$) | 77–300 | – | PTD | [22, 23] |
| | $Yb^{3+}$ | ABCYS ($AlF_3$-$BaF_2$-$CaF_2$-$YF_3$-$SrF_2$) | RT | 0.13 | IR | [24] |
| | $Tm^{3+}$ | ZBLAN ($ZrF_4$-$BaF_2$-$LaF_3$-$AlF_3$-NaF) | RT | 24 | IR | [25] |
| | | | RT | 19 | IR | [26] |
| | | | RT | 1.2 | IR | [27] |
| Fluoro-chloride | $Yb^{3+}$ | CNBZn ($CdF_2$-$CdCl_2$-NaF-$BaF_2$-$BaCl_2$-$ZnF_2$) | 77–300 | – | PTD | [22, 23] |
| | $Er^{3+}$ | CNBZn ($CdF_2$-$CdCl_2$-NaF-$BaF_2$-$BaCl_2$-$ZnF_2$) | RT | 0.5 | IR | [28, 29] |
| **Crystals** | | | | | | |
| Oxides | $Yb^{3+}$ | $KGd(WO_4)_2$ | RT | – | PTD | [30] |
| | $Yb^{3+}$ | $KY(WO_4)_2$ | RT | – | PTD | [31] |
| | $Yb^{3+}$ | YAG ($Y_3Al_5O_{12}$) | RT | 8.9 | IR | [32] |
| | $Yb^{3+}$ | $Y_2SiO_5$ | RT | 1 | IR | [32] |
| Fluorides | $Yb^{3+}$ | $BaY_2F_8$ | RT | 4 | IR | [33–35] |
| | $Yb^{3+}$ | $YLiF_4$ | RT | 69 | DLT | [36] |
| | | | RT | 6.3 | DLT | [35] |
| | $Tm^{3+}$ | $BaY_2F_8$ | RT | 1.5 | IR | [37] |
| | | | RT | 3.2 | | [38] |
| Chlorides | $Yb^{3+}$ | $KPb_2Cl_5$ | 70–300 | – | PTD | [23] |
| | | | 150–300 | – | PTD | [39] |
| | $Er^{3+}$ | $KPb_2Cl_5$ | RT | 0.7 | IR | [28, 29] |

Analysis of many commercial ZBLAN:$Yb^{3+}$ samples has shown that laser cooling efficiencies can vary by a factor of five between nominally identical samples. Similar observations have been made in other materials, indicating pronounced sensitivity to impurities and material imperfections that are not being properly controlled during the material fabrication. Recent theoretical studies of the $Yb^{3+}$ system found that specific impurities are expected to be particularly detrimental to laser cooling [7]. These include the ubiquitous hydroxyl ($OH^-$) ion, various transition-metal ions such as $Cu^{2+}$, $Fe^{2+}$, $Co^{2+}$, and $Ni^{2+}$, and microscopic defects in the glass. The presence of such impurities opens up various pathways for the excitation to decay by nonradiative relaxation (i.e. causing "parasitic" heating) rather than by $Yb^{3+}$ anti-Stokes fluorescence (i.e. causing cooling). The reduction of such impurities in the starting materials and the use of ultraclean methods to fabricate the final glass or crystal are the focus of the current material development in this field. These are the same impurities that were previously found to dominate the propagation loss in fluoride fibers, and many of the purification techniques developed in the late 1990s for fluoride fiber optics can be applied to the fabrication of high-purity laser cooling materials. Parts-per-billion-grade materials are poised to reveal the intrinsic laser cooling potential of these rare-earth-doped solids in the years to come. The goal is to realize efficient laser cooling of bulk samples to cryogenic temperatures, that is < 150 K. This temperature range is not accessible with thermoelectric coolers, and optical refrigerators may find many applications where compact, efficient, reliable, and vibration-free cooling in the 100–150 K range is needed.

## 2.2
## Material Design Considerations

Figure 2.1 illustrates the process of optical refrigeration in a system with two broadened states. Absorption occurs at a particular energy $E_P$ and is followed by anti-Stokes luminescence at a slightly higher energy, $E_F$. The energy difference $\Delta E = E_F - E_P$ is supplied via interactions with the phonons of the host material and is on the order of several $kT$. Each successful absorption–emission cycle extracts $\Delta E$ of heat from the solid on average and cools the material with an efficiency of $\eta_{cool} = (E_F - E_P)/E_P$. The performance of this ideal case is degraded in the presence of nonradiative relaxation of the excited state ($w_{mp}$), which competes with radiative relaxation ($w_r$) and introduces unwanted heating through the creation of phonons. This multiphonon relaxation reduces the quantum efficiency to $\eta = w_r/(w_r + w_{mp}) < 1$ and consequently degrades the cooling efficiency to [7]

$$\eta_{cool} = (\eta E_F - E_P)/E_P. \qquad (2.1)$$

The type of active ion and the type of material that hosts the active ion are two critical parameters that determine the quantum efficiency $\eta$ and thus the cooling efficiency $\eta_{cool}$.

### 2.2.1
### Active Ions

#### 2.2.1.1
#### Rare-Earth Ions for Laser Cooling

The active ion in a laser cooling material has the role of absorbing the light of the pump laser and re-emitting it, with high quantum efficiency, at a slightly higher energy. Rare-earth ions are well suited for realizing a system such as the one shown in Figure 2.1. The rare earths are the group of elements ranging from lanthanum to lutetium in the periodic table. The most stable oxidation state of rare earths in solids is 3+, and the respective electron configurations consist of completely filled orbitals up to xenon plus partially filled 4f orbitals. The electron configuration is denoted by [Xe]$4f^N$, where $N$ ranges from $N = 1$ for cerium ($Ce^{3+}$) to $N = 14$ for lutetium ($Lu^{3+}$). The 4f electrons are primarily responsible for the optical properties of the rare earths. The 4f wavefunctions are unique in that they are deeply buried within the electron shell of the rare-earth ion and thus shielded from the surroundings. Their small radial extent leads to negligible overlap with the orbitals of nearby atoms, causing the 4f electrons of a rare-earth ion in a solid to behave similarly to those of the ion in free space. This shielding of the 4f electrons leads to spectrally narrow optical transitions and high luminescence quantum efficiencies – both of which are necessary attributes for optical refrigeration.

A given [Xe]$4f^N$ electron configuration gives rise to a set of 4f electronic states in the presence of the electrostatic (Coulomb) interactions between the 4f electrons and the spin–orbit interaction. The resulting states are denoted by the term $^{2S+1}L_J$, where $L$, $S$, and $J$ are the orbital angular momentum, spin, and angular momentum quantum numbers, respectively [40]. The $^{2S+1}L_J$ energy levels of the trivalent rare-earth ions are shown in Figure 2.2. All rare-earth ions have a large number of 4f electronic states with the exception of cerium ($Ce^{3+}$) and ytterbium ($Yb^{3+}$). The latter electron configurations have only one 4f electron ($Ce^{3+}$) or, equivalently, lack only one 4f electron to fill the 4f shell ($Yb^{3+}$), and therefore simply produce a $^2F$ manifold that is split into the two multiplets $^2F_{7/2}$ and $^2F_{5/2}$ by spin–orbit coupling. In the following, we will focus on the ground state and the first excited state of each of these ions (indicated in bold in Figure 2.2). Using $Er^{3+}$ as the active ion, Fernandez et al. have shown that higher excited states can be used for laser cooling [28,29].

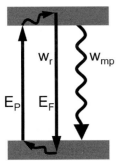

**Figure 2.1** Optical refrigeration in a system with two broadened states. $E_P$ and $E_F$ denote the pump energy and mean luminescence energy, respectively. Multiphonon relaxation of the excited state (vertical wavy arrow) competes with the desired radiative relaxation, and the respective rate constants $w_{mp}$ and $w_r$ determine the quantum yield $\eta$ and the cooling efficiency $\eta_{cool}$ [see (2.1)].

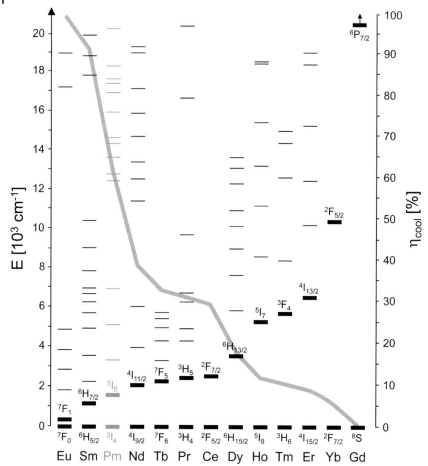

**Figure 2.2** Left axis: 4f energy levels of the trivalent rare-earth ions for energies up to 20 000 cm$^{-1}$ (adapted from [41]). The rare-earth ions are ordered with respect to increasing energy of the first excited state (marked bold). Promethium (Pm) is the only radioactive rare-earth element and is shown in gray. Right axis: Ideal cooling efficiency $\eta_{cool} = \Delta E / E_P$ ((2.1) for $\eta = 1$), assuming $\Delta E = E_F - E_P = 2.5 kT \approx 500$ cm$^{-1}$ for 300 K (gray curve).

However, the higher excited states can then relax to more than one lower-lying state and thereby introduce additional nonradiative processes that cause heating.

Each $^{2S+1}L_J$ free-ion state has a $(2J + 1)$-fold degeneracy that is partially or completely lifted under the influence of the electrostatic field produced by nearest neighbors in a solid. The type and magnitude of this crystal-field splitting is determined by the type of nearest neighbors and the symmetry of the electrostatic field they produce at the rare-earth ion site. It is this crystal-field splitting that provides the level broadening shown in Figure 2.1, thus allowing the pump energy and average emission energy to differ and thereby enabling the laser cooling cycle.

From (2.1) it follows that the cooling efficiency increases as the energy of the excited state, $E_P$, decreases. To illustrate this, the rare-earth ions shown in Figure 2.2 are ordered with respect to increasing energy of the first excited state, and the respective ideal (room temperature) cooling efficiency $\eta_{cool} = (E_F - E_P)/E_P$ is shown. Based solely on this energy argument, one would tend to choose rare-earth ions on the left side in Figure 2.2 as active ions for laser cooling because they could offer cooling efficiencies of > 30% in principle. However, both nonradiative relaxation (see Section 2.2.2.1) and the availability of suitable pump lasers limit the possible choices. Semiconductor diode lasers are attractive pump sources for integration into optical refrigerators because they can be compact, reliable, and efficient. Semiconductor lasers offering useful continuous wave (cw) output powers of several hundred mW per facet or more at room temperature are limited to a wavelength range of about 0.6–2.6 μm [42], with the highest power at around 1 μm. This range can be significantly expanded with optical parametric oscillators (OPOs). In particular, cw OPOs based on periodically poled lithium niobate (PPLN) have achieved multiwatt output power in the 1–4.5 μm wavelength range [43]. At present, no compact high-power lasers exist for wavelengths longer than 4.5 μm, where nonlinear oxide materials exhibit strong absorption. The availability of practical pump sources limits the choice of active ions to those with an energy of the first excited state that is greater than ~ 2000 cm$^{-1}$. These ions include $Yb^{3+}$, $Er^{3+}$, $Tm^{3+}$, $Ho^{3+}$, $Dy^{3+}$, and $Ce^{3+}$ (see Figure 2.2). The first excited state of $Gd^{3+}$ is in the UV spectral range and is not a practical choice due to its intrinsically low cooling efficiency.

### 2.2.1.2
**Active Ion Concentration**

To maximize the cooling efficiency, the pump laser is tuned to the longest workable wavelength in the long-wavelength tail of the absorption spectrum to make the energy difference $\Delta E = E_F - E_P$ as large as possible [see (2.1)]. A multipass pump geometry ensures high absorption of laser power when the absorption coefficient is low. The mirrors used to implement this scheme have residual absorption loss and may be a source of heating. This is especially detrimental if the mirror surfaces are deposited directly on the cooling element. As a result, there is a tradeoff between maximizing the pump wavelength, minimizing the number of mirror reflections, and keeping the sample volume as small as possible. The situation is improved by increasing the pump absorption coefficient via the active ion concentration. As more active ions are added, however, their average separation decreases, leading to more ion–ion interactions. Such interactions can mediate nonradiative transfer of energy from an excited active ion to another nearby active ion; the energy can migrate in the active ion sublattice if the ion concentration is sufficiently high. Energy transfer between two similar ions has a high degree of resonance and, on average, does not cause unwanted heating. Energy migration however increases the probability that the excitation will reach an impurity site where it decays nonradiatively, causing heating. This so-called *concentration quenching* sets an upper limit for the active ion concentration, depending on the level of purity of the host material. The optimum choice of pump wavelength and active ion concentration, therefore, is a complicated tradeoff governed by the quality of both the laser cooling material and the mirrors.

### 2.2.2
### Host Materials

A number of material characteristics are important when choosing a laser cooling host. The primary parameter is the energy of the highest-energy optical phonon, which determines the rate of multiphonon relaxation of rare-earth ions and thus, in turn, the quantum efficiency and the laser cooling efficiency. Secondary parameters include chemical durability, mechanical properties, thermal properties, and refractive index.

#### 2.2.2.1
#### Multiphonon Relaxation

The relaxation of an excited state is a competition between radiative relaxation (emission of a photon) and multiphonon relaxation (emission of phonons). The presence of multiphonon relaxation reduces the quantum efficiency $\eta$ and degrades the cooling efficiency $\eta_{cool}$ [(2.1)]. The transition between the ground state and the first excited state of rare-earth ions has an oscillator strength of typically $\sim 10^{-6}$ [44] and, therefore, the radiative relaxation rates, $w_r$, of the first excited states are on the order of $10^2$ s$^{-1}$ [45]. The host material has to be chosen such that the excited state of the active ion has a multiphonon relaxation rate, $w_{mp}$, that is significantly smaller than the radiative relaxation rate, $w_r$, which makes the quantum efficiency $\eta = w_r/(w_r + w_{mp})$ close to 1. The rate $w_{mp}$ depends primarily on the number of phonons $m = E_P/\hbar\omega$ that are created during the relaxation, where $\hbar\omega$ is the energy of the accepting phonon mode. This behavior is described by the well-known energy-gap law [46]

$$w_{mp} = \beta e^{-\alpha m}, \qquad (2.2)$$

where $\alpha$ and $\beta$ are constants that are characteristic of the host material. Of the many acoustic and optical phonons present in a material, it is the highest-energy vibration $\hbar\omega_{max}$, typically a localized optical mode, that is most likely to be the accepting mode in a multiphonon relaxation. This mode creates the lowest number of phonons and, according to (2.2), has the highest rate.

One can now estimate the maximum phonon energy, $\hbar\omega_{max}$, that produces a multiphonon relaxation rate that is still acceptably small compared to $w_r$, which does not substantially lower the quantum efficiency and degrade the cooling efficiency [(2.1)]. To do so, let us assume that (1) the typical radiative relaxation rate of the first exited state is $w_r \approx 10^2$ s$^{-1}$, and that (2) a material realizing > 90% of the theoretically possible cooling efficiency is acceptable. Based on values from

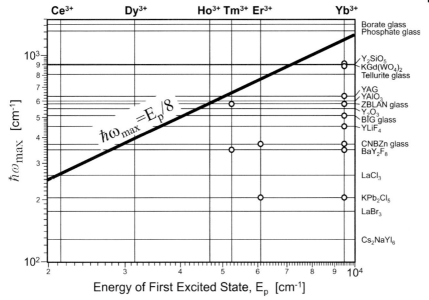

**Figure 2.3** Combinations of active ions (Section 2.2.1) and host materials (Section 2.2.2) for optical refrigeration. Combinations for which the energy of the highest-energy optical phonon, $\hbar\omega_{max}$, is less than $E_P/8$ are expected to achieve > 90% of the ideal cooling efficiency $\eta_{cool} = (E_F - E_P)/E_P$. Materials in which laser cooling has been experimentally observed (see Table 2.1) are indicated by the open circles. Values for $\hbar\omega_{max}$ were taken from [35, 47, 48]. Values for $E_P$ were calculated by subtracting $\Delta E = E_F - E_P \approx 2.5kT \approx 500\ cm^{-1}$ (at 300 K) from the absorption baricenter energies reported in [49].

a wide range of materials [47], one finds that efficient laser cooling is possible if $\hbar\omega_{max} < E_P/8$, a criterion by which useful combinations of host materials and active ions can now be identified. Figure 2.3 illustrates this criterion for the candidate active ions $Yb^{3+}$, $Er^{3+}$, $Tm^{3+}$, $Ho^{3+}$, $Dy^{3+}$, and $Ce^{3+}$ (identified in Section 2.2.1) and various host materials. Note that all of the materials showing cooling to date (Table 2.1) fall below the $\hbar\omega_{max} < E_P/8$ line, confirming the validity of this simple criterion. Clearly, fluoride glasses and fluoride crystals are ideal host materials due to their low phonon energies compared to oxide glasses. Oxide crystals such as YAG, YAlO$_3$, and Y$_2$O$_3$ have phonon energies similar to ZBLAN glass and are possible host materials for laser cooling when doped with $Yb^{3+}$, $Er^{3+}$, and $Tm^{3+}$. Phonon energies of silica-based oxide glasses are generally too high to realize laser cooling, while some oxide glasses such as tellurides with lower energy phonons may work with $Yb^{3+}$. Laser cooling with ions that have low-energy first excited states, such as $Dy^{3+}$ or even $Ce^{3+}$, is attractive because high room-temperature cooling efficiencies of ~ 17% and ~ 29%, respectively, are theoretically possible. Laser cooling may be possible in BaY$_2$F$_8$:Dy$^{3+}$, while laser cooling with $Ce^{3+}$ is predicted to require a chloride or bromide host material such as LaCl$_3$, LaBr$_3$, or KPb$_2$Cl$_5$ for multiphonon relaxation of the $^2F_{7/2}$ excited state to be sufficiently low.

#### 2.2.2.2
**Chemical Durability**

The chemical durability of a laser cooling material dictates the material growth processes, the subsequent sample processing techniques, and the ultimate material reliability of a device. The primary concern is degradation via interactions with water, which can occur during wet-chemical cutting and polishing as well as during long-term exposure to a humid atmosphere. Oxide crystals, oxide glasses, and fluoride crystals are generally stable in humid environments and greatly facilitate handling procedures for such materials. In contrast, fluoride glasses have a tendency to corrode in liquid water, and all heavy halide (chloride, bromide, iodide) compounds are extremely prone to hydrolysis. This hygroscopicity adds complexity to the material handling and device integration, and restricts the application potential in certain cases, especially when the surface-to-volume ratio is large, such as in long optical fibers or thin films. Applications of hygroscopic materials in bulk form are however possible, as shown, for example, by the recent successful commercialization of very hygroscopic $Ce^{3+}$-doped lanthanum bromide and lanthanum chloride crystals as gamma-ray scintillators.

Heavy-metal fluoride glasses react rapidly with liquid water at ambient temperature. Fluorozirconate glasses such as ZBLAN are particularly prone to hydrolysis in liquid water, while fluoroaluminate and fluoroindate glasses are less susceptible. Reaction with gaseous water, on the other hand, occurs at a very slow rate at ambient temperature [50]. Water can attack the fluoride glass surface and break bridging fluoride bonds, producing Zr–O–Zr and Zr–OH surface species [51, 52]. In these early stages of glass corrosion, a corroded surface layer forms consisting primarily of zirconium hydroxide, zirconium oxyfluoride, and various barium and lanthanum fluoride species. Hydroxides are particularly detrimental to laser cooling, since the high vibrational energy of the O–H bond enhances multiphonon relaxation of nearby excited active ions and thus degrades the laser cooling efficiency. The resulting surface heating may be pronounced and can deteriorate the performance of the entire device. These problems can be minimized by excluding water from the ZBLAN fabrication processes. Furthermore, thin-film coatings such as $SiO_2$, BN, TiN, $Si_3N_4$, TiC, SiC, $TiB_2$, $B_4C$, $MgF_2$, $TiSi_2$, Ti, $ThF_4$, and MgO have been studied extensively to protect ZBLAN fibers from hydrolysis [53–55]. For example, a 300 Å thick layer of MgO was shown to protect ZBLAN from attack by liquid water (20 °C for 100 h) [53]. MgO is particularly attractive since it is easy to deposit, is transparent in the wavelength range relevant for laser cooling, and has favorable adhesion properties. Such a thin-film coating could ensure the long-term reliability of the ZBLAN cooling element in an optical cryocooler.

#### 2.2.2.3
**Thermal and Thermomechanical Properties**

The thermal and thermomechanical properties of a laser cooling material are important parameters for the integration of the various components of an optical cryocooler. Examples of the thermal expansion coefficients, thermal conductivities, and heat capacities of some laser cooling materials are given in Table 2.2.

**Table 2.2** Coefficients of thermal expansion $\alpha$, thermal conductivities $\kappa$, and heat capacities $C$ of various laser cooling materials at room temperature and 100 K. The data are compiled from [56–58].

| | $\alpha$ (300 K) [$10^{-6}$ K$^{-1}$] | $\alpha$ (100 K) [$10^{-6}$ K$^{-1}$] | $\kappa$ (300 K) [W m$^{-1}$K$^{-1}$] | $\kappa$ (100 K) [W m$^{-1}$K$^{-1}$] | $C$ (100 K) [J g$^{-1}$K$^{-1}$] |
|---|---|---|---|---|---|
| ZBLAN | 27.5 | – | 0.76 | – | 0.67 |
| YLiF$_4$ | 10.05 (c-axis) 14.31 (a-axis) | 3.18 (c-axis) 2.36 (a-axis) | 4.1 (a-axis) 5.2 (c-axis) | 11.3 (a-axis) 13.7 (c-axis) | 0.79 |
| BaY$_2$F$_8$ | 17.2 (b-axis) | 9.5 (b-axis) | 3.5 (b-axis) | 9.7 (b-axis) | 2.83 |
| YAG | 6.14 | 1.95 | 11.2 | 46.1 | 0.59 |

The laser cooling element must be placed between mirrors to realize a multipass geometry for the pump light. Thiede et al. used two dielectric thin-film mirrors deposited directly onto the parallel surfaces of a ZBLAN cylinder to confine the pump light that was launched through a pinhole in one of the mirrors [9], while Seletskiy et al. used two external mirrors to couple the light into the cooling element by means of a cavity-enhanced resonant absorption scheme [36]. Mirrors deposited directly onto the surface of the laser cooling material will cool along with the cooling element. This requires good adhesion of the dielectric thin-film mirrors to the cooling element over the entire range of operating temperatures. Delamination of the film can occur if the shear stresses at the film-substrate interfaces exceed adhesive forces. This can be caused by substantially different thermal expansion coefficients of the cooling-element substrate and the thin film.

The rate of heat transfer through a material is proportional to the temperature gradient and the thermal conductivity. In an optical refrigerator, heat will flow from a heat load through a thermal link into the laser cooling element, where it is carried away as photons by the laser cooling cycle shown in Figure 2.1. High thermal conductivities of both the thermal link and the cooling element itself will allow the optical refrigerator (1) to cool quickly from room temperature to its cryogenic operating temperature, and (2) to react quickly to changing thermal loads during operation. The thermal conductivity of ZBLAN is quite low compared to crystalline materials, which have an advantage in this respect (see Table 2.2).

#### 2.2.2.4
**Refractive Index**

The index of refraction affects a laser cooling material in two quite distinct and counteracting ways.

First, the refractive index governs the escape of fluorescence light. A typical application will likely use a highly symmetric cooling element, such as a cylinder or a cube, and the escape efficiency for light emitted inside the element is the result of the interplay between refraction, total internal reflection (TIR), and absorption. The inset in Figure 2.4(left) shows that rays can be (1) refracted at the first surface, (2) refracted after several TIR, or (3) remain trapped indefinitely by TIR. Total inter-

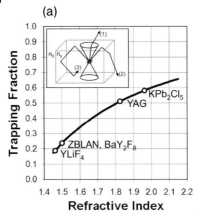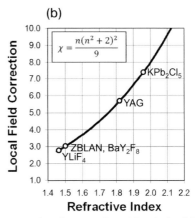

**Figure 2.4** *Left*: Refraction and total internal reflection (TIR) of isotropically emitted light inside a rectangular parallelepiped sample shape. The *inset* shows rays that are (1) refracted at the first surface, (2) refracted after several TIRs, or (3) remain trapped indefinitely by TIR. The *curve* shows the fraction of light trapped in this geometry as a function of sample refractive index. *Right*: The local-field correction term, $\chi$, for electric-dipole induced radiative relaxation as a function of refractive index. The radiative relaxation rate scales linearly with $\chi$. Refractive indices for the exemplary materials in both graphs are for a wavelength of ~ 1 μm.

nal reflection increases the average path length for the fluorescence light to escape the sample and, as a consequence, increases the probability of reabsorption by an impurity, a process that will cause unwanted heating. The fraction of light emitted into directions that remain indefinitely trapped depends on both the sample geometry, the refractive index of the sample, and the refractive index of the surrounding medium, $n_0$ [59]. A perfect rectangular parallelepiped, for example, will indefinitely trap a fraction $F = 3\cos(\phi_c) - 2$ of the emitted light inside the body, where $\sin\varphi_c = n_0/n_s$ defines the critical angle of TIR, and $n_s \geq \sqrt{2}$. This fraction is shown in Figure 2.4(left), and it becomes substantial for materials with refractive indices of > 1.6. If necessary, this effect can be reduced by deliberately using a cooling element with lower symmetry.

Second, the rate for spontaneous radiative relaxation of an excited state scales linearly with the local-field correction factor, $\chi = n(n^2 + 2)^2/9$, which is a function of the refractive index $n$ at the emission wavelength [60]. As shown in Figure 2.4 (right), the local-field correction varies by a factor of ~ 3 for typical laser cooling materials. The higher radiative rates in high-index materials enhance the quantum efficiency and thus the cooling efficiency. While this effect is small, it can become important at lower temperatures where $\Delta E = E_F - E_P$ is significantly smaller than at room temperature and the cooling efficiency becomes more sensitive to the quantum efficiency [see (2.1)]. It is conceivable that the advantage of a large local-field correction factor outweighs the disadvantage of enhanced reabsorption by TIR at high refractive indices, especially if the material is of high purity and reabsorption by impurities is small.

## 2.2.3
### Material Purity

Impurities in the context of optical refrigeration are defined as any species that can induce nonradiative relaxation and thus produce unwanted heating. The first class of impurities includes macroscopic material defects such as foreign particulates and precipitates that absorb at the pump wavelength and relax nonradiatively. Most of these defects can be minimized by ensuring careful control of equipment and environmental cleanliness during fabrication. Each material will require a specifically developed procedure. For example, the various fluoride constituents in a fluoride glass melt dissolve at different rates. Insufficient melting times can leave residual undissolved particles such as $YbF_3$ that act as scattering centers or absorbers and can cause heating. Fluorozirconate glasses such as ZBLAN have $ZrF_4$ as their main component, which has a pronounced tendency to reduce to lower oxidation states of zirconium during the melting in a reducing atmosphere [61, 62]. The reduction of $ZrF_4$ causes the formation of black precipitates that heat when excited in the near-infrared spectral region, strongly reducing the laser cooling efficiency of ZBLAN. Reduction of $ZrF_4$ and the respective formation of black precipitates can be suppressed by carefully adjusting the concentration of an additional oxidizer component such as $InF_3$ [63, 64] (see Section 2.4.1).

A second class of impurities includes species that can quench the active ion excited state at the atomic level. The main processes are illustrated in Figure 2.5 and are described in the following.

#### 2.2.3.1
**Vibrational Impurities**

Multiphonon relaxation can occur via interactions with the phonons of the host material, as described in Section 2.2.2.1. Multiphonon relaxation can however also occur via interactions with molecular species that are incorporated into the material at trace levels. The main culprits are the hydroxyl ion ($OH^-$) and mixed ox-

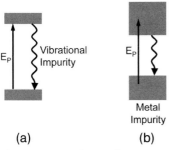
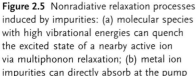
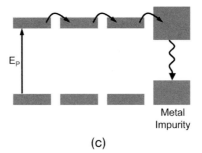

**Figure 2.5** Nonradiative relaxation processes induced by impurities: (a) molecular species with high vibrational energies can quench the excited state of a nearby active ion via multiphonon relaxation; (b) metal ion impurities can directly absorb at the pump wavelength and subsequently relax nonradiatively; (c) excitation energy can migrate by resonant nonradiative energy transfer among active ions and reach a metal ion impurity, where it relaxes nonradiatively.

ides (for example oxyfluorides and hydroxyfluorides), which can form from residual water during the glass preparation or crystal growth process [65]. Other problematic species include ammonium ($NH_4^+$), sulfate ($SO_4^{2-}$), phosphate ($PO_4^{3-}$), and carbon dioxide ($CO_2$). They all have vibrational modes in the near-infrared spectral region [65–70] that can act as acceptors in multiphonon relaxation processes, as illustrated in Figure 2.5a. For example, a multiphonon relaxation involving the 3440 cm$^{-1}$ OH$^-$ stretching frequency [69,71] as an accepting mode creates only 1–3 phonons when de-exciting the first excited states of the laser cooling ions shown in Figure 2.3. For $m < 3$, multiphonon relaxation dominates over radiative relaxation by orders of magnitude. The quantum efficiency of any such ion with an OH$^-$ impurity nearby will be essentially zero, greatly deteriorating the laser cooling efficiency. Model calculations indicate that even for $Yb^{3+}$, which has a large energy gap between the ground and first excited states, the concentration of the OH$^-$ impurity has to be less than ~ 100 ppb for efficient optical refrigeration to be possible at cryogenic temperatures [7]. Specific processing steps are required during fabrication to ensure that the concentration of such vibrational impurities in the final product meets these challenging purity requirements.

#### 2.2.3.2
**Metal-Ion Impurities**

Many metal ions have strong absorption in the visible and near-infrared wavelength ranges and, more often than not, these ions have a low quantum efficiency. Such metal-ion impurities can adversely affect the laser cooling efficiency in two distinct ways. First, metal-ion impurities can directly absorb at the pump wavelength and subsequently relax nonradiatively, as illustrated in Figure 2.5b. This effect is referred to as *background absorption*. Second, excitation energy can reach metal-ion impurities via the active ions. As the active ion concentration is increased to percent levels, nonradiative energy transfer between active ions becomes competitive with radiative relaxation and so the excitation energy can migrate among the active ions. If a metal-ion impurity is present within the energy-migration volume there is a chance of transferring the excitation energy to this impurity, which will subsequently relax nonradiatively, as illustrated in Figure 2.5c. This effect is referred to as *concentration quenching*. Metal ions that absorb at laser cooling wavelengths can therefore degrade the cooling efficiency by both background absorption (spectral overlap with pump laser) and concentration quenching (spectral overlap with active-ion emission). Absorption spectra of common transition metal ions are shown in Figure 2.6, along with the approximate wavelengths of the transitions of the rare-earth laser cooling ions discussed in Section 2.2.1. Of highest concern are the divalent transition-metal ions $Cu^{2+}$, $Fe^{2+}$, $Co^{2+}$, and $Ni^{2+}$, which have strong and spectrally broad absorptions in the 1 – 3 μm wavelength range. Moreover, these ions have a high natural abundance and are common impurities in a wide range of chemicals. The impact of a given metal-ion impurity on the cooling efficiency depends on the spectral overlap between the impurity absorption and the active-ion emission. The laser cooling efficiencies of $Yb^{3+}$, $Er^{3+}$, $Tm^{3+}$ and $Ho^{3+}$ are expected to be more sensitive to these divalent transition-metal impurities than $Dy^{3+}$ and

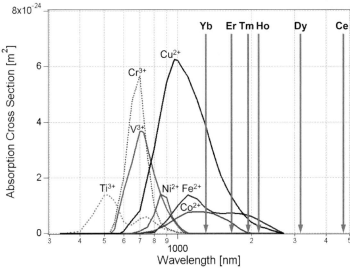

**Figure 2.6** Absorption cross-sections for various transition-metal ions in ZBLAN glass at room temperature (adapted from France et al. [72]). The *vertical arrows* indicate the approximate wavelengths of the transitions of the rare-earth laser cooling ions discussed in Section 2.2.1.

$Ce^{3+}$. In fact, the latter two active ions may have a fairly high tolerance for metal-ion impurities which, along with their intrinsically high cooling efficiencies, makes them attractive for future studies (see Section 2.6.3). For the common $Yb^{3+}$ ion to provide practical laser cooling at cryogenic temperatures, it is estimated that the concentration of these divalent transition metal ions must be reduced to well below 100 ppb and to below 2 ppb in the case of $Cu^{2+}$ [7]. Such trace-metal concentrations exceed the purity limits of typical commercial chemicals. Targeted purification processes during the fabrication of the laser cooling material are therefore required.

Another potential source of heating arises from rare-earth impurity ions other than the active ion itself. The ion combinations $Yb^{3+}$-$Tm^{3+}$ and $Yb^{3+}$-$Er^{3+}$ are of particular concern because efficient nonradiative energy transfer can occur within these pairs [73]. Both $Er^{3+}$ and $Tm^{3+}$ can be promoted to highly excited states via either multiple sequential energy transfers from $Yb^{3+}$ (in a process known as energy-transfer upconversion) or excited state absorption. The highly excited states can subsequently relax by either radiative decay or multiphonon relaxation. Goldner et al. have studied the effect of $Er^{3+}$ and $Tm^{3+}$ impurities in ZBLAN on the cooling efficiency of $Yb^{3+}$ [74]. For a $Yb^{3+}$ concentration of $5 \times 10^{20}$ ion/cm$^3$, they found that the effect on the $Yb^{3+}$ quantum efficiency was negligible if the $Er^{3+}$ and $Tm^{3+}$ impurity concentrations did not exceed 5 and 500 ppm, respectively. These impurity threshold concentrations are more than three orders of magnitude higher than those of $Cu^{2+}$, $Fe^{2+}$, $Co^{2+}$, and $Ni^{2+}$. Purification methods targeted at removing rare-earth impurities from high-purity commercial starting materials are therefore not likely needed.

## 2.3
### Preparation of High-Purity Precursors

The material design considerations presented in Section 2.2 indicated that fluorides are one of the most attractive material classes for optical refrigeration. Fluorides offer low phonon energies to enable laser cooling with $Yb^{3+}$, $Tm^{3+}$, $Er^{3+}$, and possibly $Ho^{3+}$ (see Figure 2.3), and they are of sufficient chemical durability for integration into reliable devices. Not surprisingly, of the 15 materials that have cooled to date, nine are fluoride-based compounds. The following sections on the preparation of precursor materials and the fabrication of glasses and crystals are therefore primarily drawn from experience with fluorides. ZBLAN glass takes center stage since it is the most thoroughly explored laser cooling material to date. However, most of the ZBLAN chemistry presented in the following applies directly to the preparation of other fluoride glasses and crystals. This section should also provide useful guidance for the fabrication of oxide, chloride, and bromide materials.

### 2.3.1
### Strategies for Preparing High-Purity Precursors

Laser cooling materials require an exceedingly high purity in order to maximize the quantum efficiency of the active ion and thus maximize the laser cooling efficiency. Section 2.2.3 explained that metal-ion impurities such as $Cu^{2+}$, $Fe^{2+}$, $Co^{2+}$, $Ni^{2+}$, and $OH^-$ must not exceed low ppb levels in systems using $Yb^{3+}$ as the active ion. Consequently, all of the precursor materials used in the preparation of a laser cooling material have to meet or exceed this level of purity. Currently available commercial chemicals are usually of lower purity. In fact, ZBLAN glass fabricated directly from off-the-shelf commercial metal fluoride precursors usually shows laser-induced heating. The problem is compounded by the fact that most metal fluorides have a low solubility in water or acids and cannot be sublimated ($ZrF_4$ being a noteworthy exception). This limits the available purification techniques and necessitates alternative purification strategies.

Attempts were made in the past to remove metal-ion impurities directly from ZBLAN glass melts using electrochemical processes (see Section 2.3.3.5). In contrast, today's strategies focus on the preparation of the *individual* metal fluoride precursors in high purity, and on the subsequent preparation of the laser cooling material from these precursors. Figure 2.7 illustrates two possible strategies for the synthesis of high-purity fluorides. One approach is via an aqueous route: a high-purity inorganic precursor is dissolved in water or acid, the solution undergoes purification, the metal fluoride is precipitated with hydrofluoric acid, and the anhydrous metal fluoride is then obtained by drying the precipitated fluoride in hydrogen fluoride (HF) gas at elevated temperature. Fluorination of metal oxides in HF gas has been applied by a number of workers [75–80] using a variety of process parameters (see Section 2.3.3.3). The aqueous phase

| AQUEOUS ROUTE | | | | | | | | |
|---|---|---|---|---|---|---|---|---|
| Precursor | → | Aqueous Solution | → | Purification | → | Precipitation of Fluoride | → | Drying in HF Gas |
| Zirconium oxychloride $ZrOCl_2 \cdot 8H_2O$ | | $Zr^{4+}$ | | Filtration and Solvent Extraction | | $ZrF_4 \cdot H_2O$ | | $ZrF_4$ |

| SOL-GEL ROUTE | | | | | | | | |
|---|---|---|---|---|---|---|---|---|
| Precursor | → | Purification | → | Hydrolysis and Polycondensation | → | Gelation and Drying | → | Fluorination and Drying in HF Gas |
| Zirconium tert-butoxide $Zr[OC(CH_3)_3]_4$ | | Distillation | | -Zr-O-Zr- -Zr-OH-Zr- | | $ZrO_2 + H_2O$ | | $ZrF_4$ |

**Figure 2.7** Aqueous and sol–gel synthesis routes to anhydrous metal fluorides. Zirconium is shown as an example for each approach.

can be purified by filtration, recrystallization, solvent extraction, ion exchange, or a combination of these techniques. Another approach is via a sol–gel route: a high-purity organometallic precursor such as a metal alkoxide is purified, converted to a hydrated metal oxide in the sol–gel process [81], and the anhydrous metal fluoride is obtained by fluorinating and drying the hydrated oxide in HF gas [77, 82]. The metal alkoxide precursor can be purified by distillation, sublimation, or recrystallization [83]. The advantage of the aqueous route is that it offers purification techniques that can be targeted to very specific impurities. In contrast, the sublimation or distillation of metal alkoxide precursors in the sol–gel route is less specific and often hampered by low yields that are a result of the low thermal stability of many of these organometallic compounds [83].

The approaches shown in Figure 2.7 involve the use of hydrofluoric acid and/or hydrogen fluoride gas, and a note of caution is important. Both hydrofluoric acid and hydrogen fluoride gas have to be handled with extreme care. HF gas converts to HF acid upon contact with water or the mucous membranes when inhaled. HF acid is a corrosive, but it derives additional toxicity from its ability to easily penetrate tissue and precipitate body calcium and magnesium as the respective insoluble fluoride salts. The resulting hypocalcemia and hypomagnesemia can be fatal when as little as a few percent of the body surface area is exposed to concentrated HF. The initial contact may be painless and go unnoticed, causing delayed symptoms and enhanced injury. Experiments with HF should therefore only be conducted by qualified persons who have been properly trained to work with this acid in a laboratory setting that provides the necessary engineering and administrative controls, proper personal protective equipment, and available first-aid measures.

## 2.3.2
## Process Conditions

### 2.3.2.1
### Purity of Commercial Precursors

Commercial metal fluorides have typical cationic purities of around 99.9 to 99.99%. A fraction of the corresponding 100 to 1000 ppm impurities are contaminations that have no optical transitions near typical laser cooling pump wavelengths and are, therefore, benign in the context of optical refrigeration. Examples of such benign impurities are the fluorides of the alkali (Li, Na, K, Rb, Cs) or alkaline earth (Mg, Ca, Sr, Ba) elements. The typical ppm-level presence of impurities such as $Cu^{2+}$, $Fe^{2+}$, $Co^{2+}$, and $Ni^{2+}$ is problematic, however, and prevents laser cooling materials prepared from commercial fluorides from reaching cryogenic temperatures. Taking either an aqueous or a sol–gel route for the preparation of individual metal fluorides can exploit commercial nonfluoride high-purity precursors. The commercial precursors shown in Table 2.3 for the aqueous route have total impurity levels of < 30 ppm and are good starting points for purification processes that realize a 1000-fold purification to ppb-level impurities (see Section 2.3.3). Both the aqueous and the sol–gel processing routes use solvents and acids in the course of the preparation. The use of double-distilled trace-metal-certified commercial acids and solvents is required for this purpose.

### 2.3.2.2
### Process Equipment

Special attention must be paid to the type and cleanliness of all process equipment, ranging from weighing paper and spatulas to reaction vessels and storage containers. The aqueous route uses highly corrosive acids such hydrofluoric, hydrochloric, and nitric acid at elevated temperatures. Corrosion of reaction vessels and leaching of contaminants from the container material into the batch is of particular concern. Components made of perfluoroalkoxy copolymer (PFA) resin are excellent choices for these corrosive conditions. They have a high degree of inertness, even under these harsh conditions, and can be used successfully up to temperatures of 190 °C. Containers and utensils should be leached in 1 vol % trace-metal-grade nitric acid at 60 °C for several days and thoroughly washed in ultrahigh-purity water before each use to minimize metal contaminations by the container.

Table 2.3 Examples of commercial precursors for the preparation of metal fluorides.

| Metal ion | Aqueous route | Sol–gel route [77] |
|---|---|---|
| Zr | $ZrOCl_2 \cdot 8H_2O$ | Zr n-propoxide |
| Ba | $BaCO_3$ | Ba methoxide |
| La | $La_2O_3$ | La propoxide |
| Al | $AlCl_3 \cdot 6H_2O$ | Al methoxide |
| Na | $Na_2CO_3$ | Na methoxide |
| In | $InCl_3$ | |
| Rare-earth (RE) | $RE_2O_3$ | |

### 2.3.2.3
### Clean Environment

Open-batch processing cannot always be avoided and must be carried out in a clean environment in order to minimize contamination by airborne particulates. Airborne mineral particles are a significant potential source of contamination of chemicals by metal ions during open-batch processing. Mineral dust particles consist primarily of silicon, aluminum, and iron compounds. African dust, for example, contains ~ 9 wt % of iron in the form of iron oxide (hematite, goethite) and clays (illite, kaolinite, smectite) [84]. The mass concentration of dust particles on continents in the northern hemisphere ranges from 10–1000 µg/m$^3$ of air and can vary significantly with location and season. Aerosol dust particles range in diameter from 1 nm to tens of µm. Particles with a diameter of < 10 µm have gravitational settling velocities of < 1 cm/s [85] and easily remain aerosolized, accounting for up to half of the aerosol mass. Typical aerosolized iron concentrations can therefore reach several µg/m$^3$ of air. As a result, the contamination of a gram-sized open batch by just a few µm-sized dust particles can introduce ppb-level iron impurities that are expected to significantly degrade the cooling efficiency of a material. A clean environment certified to at least Class 100 (ISO 5), defined as having < 3520 particles (≥ 0.5 µm) per m$^3$ of air, has been found adequate for this purpose. Maintaining such a low level of airborne particles during regular work requires operators to wear full-body clean-room garments and to adhere to strict operating procedures.

### 2.3.3
### Material Purification

#### 2.3.3.1
#### Filtration and Recrystallization

The first step in the aqueous route shown in Figure 2.7 is to dissolve the precursor. Many of the transition metals form colloids in solution. For example, $Fe^{2+}$ is partially oxidized to $Fe^{3+}$ and forms the mixed oxides $Fe_2O_3$, $Fe(OH)_2^+$ and $Fe(OH)_3$, which coagulate to polynuclear complexes that are dispersed as colloidal particles [86]. Klein et al. showed that aqueous solutions of zirconyl compounds such as $ZrOCl_2$ tend to contain significant amounts of suspended and colloidal matter [87]. They showed that a 500 ppm iron impurity could be reduced to 12 ppm by simple filtration through a 10-nm pore size filter, indicating that much of the iron impurity is associated with particulate and colloidal matter. Filtration is therefore an effective and simple method to eliminate undissolved particles as a first step in material purification.

Recrystallization has been used to purify $ZrOCl_2$. Crystals of $ZrOCl_2$ can be recrystallized by filtering the aqueous solution through a < 0.2 µm pore-size nylon filter, heating to 80 °C, adjusting the acidity to 7 N by adding HCl, and recrystallizing $ZrOCl_2$ by cooling slowly [88]. Slow crystallization tends to exclude impurities from the well-defined crystalline phase, preferentially leaving them in the remaining liquor, which is separated from the crystals by filtration. The process can be repeated several times if needed.

### 2.3.3.2
**Solvent Extraction Using Chelating Agents**

Extraction of transition-metal ions from an aqueous phase into an immiscible organic phase is a well-established separation technique. Suited chelating agents can be used to bind specifically to target metal ions in the aqueous solution. The resulting metal-chelate complexes are hydrophobic and therefore tend to accumulate in the organic phase. This method is widely used in the analytical sciences to extract and concentrate trace metals in an organic phase for subsequent detection [89–93]. Chelates can provide the targeted purification in the aqueous synthesis route shown in Figure 2.7. The binding of a chelate to transition-metal ions in the aqueous phase and the subsequent extraction of the metal-chelate complexes to an organic phase has been shown to reduce iron, copper, and nickel impurities to ppb levels [87, 94].

The ammonium pyrrolidine dithiocarbamate (APDC) chelate is of particular interest for the individual purification of the precursors needed for fluoride glasses and crystals. APDC has excellent affinity for the problematic transition metals but does not bind to any of the metals that are used for the fabrication of ZBLAN glass or fluoride crystals such as $YLiF_4$ and $BaY_2F_8$ (see Table 2.4). APDC therefore can be used as a single-chelate system that is suited for the removal of impurities like Fe, Cu, Ni, and Co from individual solutions of metal ions like Zr, Ba, Li, Al, Na, In, Y, or the rare earths. The use of a single-chelate system is attractive because it minimizes the development effort and simplifies processing. The effectiveness of a chelate is quantified by the distribution coefficient, which is the ratio of the metal-ion concentration in the organic phase and the metal-ion concentration in the aqueous phase. The distribution coefficient for a given metal ion is primarily determined by the pH and by the type of organic solvent that is used as the second immiscible phase. A large body of data is available for the APDC chelate in conjunction with methyl-isobutyl-ketone (MIBK) as the organic solvent. Distribution coefficients reported in the literature for the APDC/MIBK system vary significantly and are on the order of $10^1$–$10^3$ for Cu, Fe, Co, and Ni [89–91, 95]. A solvent extraction process should be able to reduce the concentration of these impurities by a factor of at least ~ 1000, thereby decreasing the initial impurity concentrations of a few ppm to a final concentration in the low ppb range. This means that at least three sequential extraction steps are necessary. Each extraction step involves spiking the aqueous phase with APDC solution, adding the MIBK solvent, mixing the two phases in a separatory funnel by vigorous shaking for 1 min, equilibrating the system for 10 min, and collecting the lower aqueous phase. Ling *et al.* have used the APDC/MIBK system with three or four successive extractions, and they report concentrations of < 10 ppb for Fe and < 5 ppb for Cu, Ni, and Co in an aqueous solution of $ZrOCl_2$ [96]. Using a diethyl dithiocarbamate (DDTC) chelate with $CCl_4$, Kobayashi reports concentrations of < 13 ppb for Fe and < 1 ppb for Cu, Ni, and Co in the final fluorides [94].

Some chelates show a pronounced dependence of the distribution coefficient on pH. Extraction of iron with DDTC, for example, is only effective in a narrow pH range between 2.5 and 4 [96]. The distribution coefficients in the APDC/MIBK sys-

**Table 2.4** Specificity of the ammonium pyrrolidine dithiocarbamate (APDC) chelate. APDC forms stable complexes with the elements inside the **bold outline** [89]. In particular, it targets the transition-metal ions Cu, Fe, Co and Ni, which are most problematic with respect to degrading the laser cooling efficiency in $Yb^{3+}$ and $Tm^{3+}$ systems. APDC does not form stable complexes with any of the metal ions that are constituents of ZBLAN glass or typical fluoride crystals (*shaded boxes*). The pyrrolidine dithiocarbamate molecular structure is also shown.

| H  |    |    |    |    |    |    |    |    |    |    |    |    |    |    |    |    | He |
|----|----|----|----|----|----|----|----|----|----|----|----|----|----|----|----|----|----|
| Li | Be |    |    |    |    |    |    |    |    |    |    | B  | C  | N  | O  | F  | Ne |
| Na | Mg |    |    |    |    |    |    |    |    |    |    | Al | Si | P  | S  | Cl | Ar |
| K  | Ca | Sc | Ti | V  | Cr | Mn | Fe | Co | Ni | Cu | Zn | Ga | Ge | As | Se | Br | Kr |
| Rb | Sr | Y  | Zr | Nb | Mo | Tc | Ru | Rh | Pd | Ag | Cd | In | Sn | Sb | Te | I  | Xe |
| Cs | Ba | 57-71 | Hf | Ta | W  | Re | Os | Ir | Pt | Au | Hg | Tl | Pb | Bi | Po | At | Rn |
| Fr | Ra | 89-103 | Rf | Db | Sg | Bh | Hs | Mt | Ds | Rg |    |    |    |    |    |    |    |

| La | Ce | Pr | Nd | Pm | Sm | Eu | Gd | Tb | Dy | Ho | Er | Tm | Yb | Lu |
|----|----|----|----|----|----|----|----|----|----|----|----|----|----|----|
| Ac | Th | Pa | U  | Np | Pu | Am | C  | Bk | Cf | Es | Fm | Md | No | Lr |

tem are less sensitive to pH than in the DDTC system. In particular, Cu, Fe, Co, and Ni can be extracted efficiently with APDC as long as the pH is > 2.5. Nevertheless, buffers are usually needed to adjust and maintain the pH during the extraction. The ammonium acetate/acetic acid buffer has been successfully used for solvent extraction with APDC/MIBK [90, 96]. Metal ions can be precipitated from the final purified aqueous solution as the corresponding metal fluoride by adding concentrated hydrofluoric acid. Sometimes the precipitate can be collected directly by decanting the remaining solution and thoroughly drying the precipitate on the hot plate. In other cases, evaporation of excess acid in a still constructed from PFA is necessary to effect the precipitation of the metal fluoride. The situation is different for zirconium and indium, two components of ZBLAN glass. Here, the precipitation of the slightly soluble $ZrF_4 \cdot H_2O$ and $InF_3$ competes with the precipitation of the respective hexafluoro-ammonium complexes – $(NH_4)_2ZrF_6$ and $(NH_4)_3InF_6$ – which can form with the ammonium ions introduced with the buffer. Such precipitates tend to be a mixture of both the fluoride and ammonium fluoride compounds and require the additional decomposition step

$$(NH_4)_2ZrF_6 \rightarrow ZrF_4 + 2NH_4F$$
$$(NH_4)_3InF_6 \rightarrow InF_3 + 3NH_4F$$

during which ammonium fluoride evaporates and $ZrF_4$ or $InF_3$ are formed. This decomposition can be carried out in glassy carbon crucibles and proceeds quantitatively at 370 °C in ambient atmosphere [97].

### 2.3.3.3
**Fluorination and Drying in Hydrogen Fluoride Gas**

The aqueous route illustrated in Figure 2.7 produces metal fluorides that contain residual oxides and water. The residual water can be either crystalline water, such as in the case of $ZrF_4 \cdot H_2O$, or water adhering to the surface of the material. If such "wet" fluorides were to be used in the fabrication of a fluoride crystal or glass, the residual surface and crystalline water would convert to hydroxides ($OH^-$ groups) and oxides during the melting step and cause unwanted nonradiative relaxation in the final material. Residual oxides can also enhance the tendency of the glass to crystallize during the quenching process. Therefore, residual surface water has to be removed and residual oxides have to be converted to fluorides before the metal fluorides can be used for glass or crystal fabrication. If the sol–gel route is used, all of the material is in the form of fine-grained "wet" oxides that have to be completely converted to fluorides. In either case, drying and fluorination can be accomplished by treatment in hydrogen fluoride (HF) gas at elevated temperature. Surface water is driven off by evaporation, and residual metal oxides are converted to fluorides following the general solid–gas reaction

$$M_2O_3 + 6HF \leftrightarrow 2MF_3 + 3H_2O.$$

The Gibbs free energy for fluorination at 300 °C is in the range of −400 to −600 kJ/mol for rare-earth oxides (i.e. the thermodynamic equilibrium of the above reaction is shifted towards the metal fluoride side). Kwon *et al.* have shown that micron-sized powders of $Nd_2O_3$, $CeO_2$, and $SrO$ can be completely fluorinated in HF gas at 300 °C in less than 1 h [79]. The HF gas is typically diluted to ∼ 30 vol % by using argon as a carrier gas. The result of this process is an ultradry, bright white metal-fluoride powder. Some studies report HF processing at higher temperatures, such as 550 °C [80] or even 800 °C [78]. Our research shows that fluorides processed in HF at such high temperatures always have a slightly gray color, which is due to carbon contamination originating from the corrosion of the glassy carbon components inside the reaction vessel. HF processing temperatures above 300 °C should be avoided because the resulting carbon particles will be incorporated into the glass, where they absorb the laser cooling pump light and cause undesired heating. The corrosion rate of the Inconel steel vessel also increases noticeably above 300 °C, and contamination of the metal fluoride batch by the respective corrosion products is possible.

The discussion in Section 2.3.1 emphasized the critical importance of safe operation when working with corrosive and very toxic HF gas. In particular, the use of HF gas cylinders is strongly discouraged because large quantities of pressurized HF gas greatly increase operational risk. A much safer approach is to produce HF gas *in situ* by the decomposition of potassium hydrogen fluoride ($KHF_2$) [78, 80]. The latter is an inexpensive salt that is stable at room temperature but melts at 220 °C and decomposes to hydrogen fluoride gas and potassium fluoride according to

$$KHF_2 \rightarrow KF + HF.$$

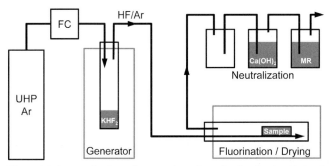

**Figure 2.8** System for the safe generation of HF gas and drying/fluorination of a metal fluoride or oxide sample. A flow controller (FC) provides a constant flow of ultrahigh-purity argon (UHP Ar) carrier gas. $KHF_2$ is decomposed in the generator furnace tube and the resulting HF/Ar gas mixture is injected into the fluorination/drying furnace tube that holds the sample at the process temperature. Excess HF gas is neutralized by bubbling it through a calcium hydroxide slurry, and a final bubbler with methyl red (MR) pH indicator reports the status of the neutralization tank and the presence of gas flow.

The rate of decomposition is a function of the changing ratio of $KHF_2$ and KF in the melt [78] and of the flow rate of the argon carrier gas. Complete and controlled decomposition can be achieved by heating 100 g of $KHF_2$ from 390 to 440 °C at a rate of 4.5 °C/h and further heating to 480 °C at a rate of 8.5 °C/h in a 150 cm³/min flow of argon carrier gas. A typical apparatus for the generation of HF and the drying/fluorination of a fluoride or oxide sample is shown in Figure 2.8. The decomposition of $KHF_2$ occurs in a glassy carbon crucible in the generator tube, and the HF gas is injected into the drying tube, where the sample is held in a glassy carbon crucible at the desired fluorination temperature. A controller maintains a constant flow rate of argon carrier gas through the system. The $KHF_2$ decomposition can be stopped quickly, if needed, by turning off the generator furnace. All hot components are constructed from Inconel Alloy 600, a nickel steel that has a high resistance to corrosion by HF. The inside of the drying tube is lined with a glassy carbon tube to prevent any steel corrosion products from contaminating the sample in the crucible. The tubing and neutralization containers are made exclusively from PFA. Residual HF gas emerging from the drying tube is bubbled through a stirred calcium hydroxide slurry, where HF is removed by precipitating it as $CaF_2$. Methyl red pH indicator is added to the final bubbler tube to indicate when the calcium hydroxide has been exhausted and the neutralization line has to be recharged. This open system operates safely at low pressure and is set up in a fume hood in case any HF gas leaks from the apparatus.

### 2.3.3.4
### Sublimation and Distillation

Sublimation of solids and distillation of liquids are common methods for separating compounds with different vapor pressures. A solid or liquid must have some minimum vapor pressure (typically around 1 Torr at a given temperature) for evaporation to be efficient and the yield of the process to be practical. The vapor pressures of most metal fluorides below 1200 K are, however, too low for physical vapor transport to be effective. Of the metal fluorides shown in Figure 2.9, only ZrF$_4$ has a vapor pressure that is sufficiently high to enable sublimation at temperatures below 1200 K. The vapor pressures of the problematic Cu, Fe, Co, and Ni fluorides are more than a factor of 1000 lower than that of ZrF$_4$, and separation of ZrF$_4$ from these impurities can be achieved by sublimation. MacFarlane et al. have shown that sublimation of ZrF$_4$ around 1135 K reduces transition-metal impurities by a factor of about two from their original levels [106]. With typical ppm-level initial transition-metal concentrations in commercial ZrF$_4$, sublimation alone is insufficient to reduce such impurities to the low ppb range needed for optical refrigeration. MacFarlane et al. found that distillation of ZrF$_4$ from a melt initially containing 81 mol % ZrF$_4$ and 19 mol % BaF$_2$ was more effective than sublimation. They achieved a reduction in transition-metal impurities by a factor of 8–10 from their original levels. ZrF$_4$-BaF$_2$ melt distillation is advantageous because (1) the vapor pressures of Fe$^{3+}$ and Fe$^{2+}$ are reduced as a result of Raoult's law and (2) some of the iron is converted to Fe$_2$O$_3$ (which has a very low vapor pressure) by the reaction of FeF$_3$ with ZrO$_2$. NaF, LiF, and AlF$_3$ can be sublimated at higher temperatures, but their vapor pressures are comparable to those of the transition-metal fluorides. Sublimation is not an effective method of purification in these cases. At 1200 K and above, the transition-metal fluorides develop a substantial vapor pressure that is at least 1000 times greater than that of BaF$_2$, AlF$_3$, YF$_3$, and the rare-earth fluorides (see Figure 2.9), making it possible to evaporate transition-metal impurities from these fluorides.

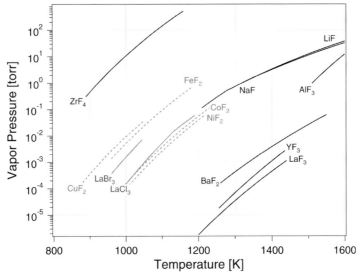

**Figure 2.9** Vapor pressures of select metal halides. Fluorides are shown in *black lines*, chlorides and bromides are shown in *gray solid lines*, and transition metal fluorides are shown in *gray dashed lines*. The data were compiled from [98–105].

### 2.3.3.5
**Electrochemical Purification**

Electrochemical purification uses an electrical current through a melt to deposit impurity metal ions in their metallic form on the electrode. This method can be applied if the electrical potential for the reduction of the impurity metal ion to the metal is much lower than that for the melt components. This is the case for the main components of ZBLAN glass melts. The exceptions are oxidizers such as $InF_3$ and $SnF_4$ that are often added to ZBLAN melts in order to prevent the reduction of $Zr^{4+}$ to lower oxidation states and to avoid the resulting formation of black $ZrF_3$ precipitates (see Section 2.4.1). Such in situ oxidizers have deposition potentials that are comparable to those of $Fe^{2+}$, $Cu^{2+}$ and $Ni^{2+}$, and therefore interfere with the electrochemical purification process. However, Bao et al. have shown that electrochemical purification can also remove black $ZrF_3$ deposits and could eliminate the need for in situ oxidizers altogether [107]. Experiments by Fajardo et al. have shown that $Yb^{3+}$-doped ZBLAN glass fabricated from ultrahigh-purity commercial metal fluorides do not show laser cooling, while the same glass treated by the electrochemical process during melting shows measurable laser cooling [108]. The improvement was ascribed to a reduction of $Fe^{2+}$ and $Cu^{2+}$ impurities. It was postulated that the lower concentration limit of an impurity realized by such an electrochemical process is determined by the dynamics of formation and dissolution of the respective metal nuclei at the electrode surface. For the problematic $Cu^{2+}$, in particular, this lower limit is around 250–300 ppb [107], a concentration that is 10–100 times above the target concentration for this impurity. While electrochemical purification can remove impurities from highly contaminated melts, this limit may not be low enough to reach the purity needed for efficient laser cooling.

### 2.3.4
**Determination of Trace Impurity Levels**

The measurement of ppb-level impurity concentrations is challenging. Trace metals can be measured directly by graphite furnace atomic absorption spectroscopy (GFAAS) or inductively coupled plasma mass spectroscopy (ICP-MS). GFAAS is capable of of low parts-per-billion (ppb) detection limits, while ICP-MS can achieve parts-per-trillion (ppt) to parts-per-quadrillion (ppq) detection limits. The analysis of impurities in a metal fluoride compound or a final fluoride glass or crystal requires digestion of the solid and the preparation of an aqueous solution that can be injected into the GFAAS or ICP-MS equipment. This has proven to be challenging in the case of fluorides because they are difficult to digest and the matrix often significantly interferes with the detection process. Newman et al. have developed a method that dissolves ZBLAN glass in a mixture of hot boric acid, nitric acid, and hydrofluoric acid, adjusts the pH of the resulting solution with acetate buffer, complexes the transition-metal ions with NaDDC and APDC chelating agents, extracts them into an organic solvent (MIBK), and back-extracts the complexes into an aqueous phase for subsequent GFAAS analysis [109]. A similar method was described earlier by Bertrand et al. [110]. While they achieved detection limits of 15, 5, and 10 ppb for Fe, Ni, and Cu, respectively, this method is prone to contamination by the many chemicals used in the sample preparation process, and it does not easily lend itself to the routine characterization of materials. Alternatively, GFAAS or ICP-MS techniques can be applied to the various intermediate steps of the aqueous preparation route discussed in Section 2.3.1 (see Figure 2.7). Measurement of trace metal levels in the final aqueous solution just before precipitation of the metal fluoride, for example, can be used as an indicator of the ultimate quality of the material. If solvent extraction with chelating agents is used, this final solution typically contains an acetate buffer at concentrations that tend to interfere with the analytical equipment. Dilution by a factor of 100 is often necessary to eliminate these matrix effects, and ICP-MS (with its lower detection limit compared to GFAAS) is therefore the method of choice.

The hydroxyl ion ($OH^-$) can be detected via optical absorption spectroscopy of the O–H fundamental stretching mode around 2.9 µm. This absorption is weak in fluoride glasses. For example, ZBLAN glass with low $OH^-$ content has an $OH^-$ absorption coefficient of only 20 dB/km, which corresponds to a loss of only ~ 5 × $10^{-5}$ $cm^{-1}$ [71]. An accurate measurement of the $OH^-$ concentration requires tens of meters of path length, making this method useful for the characterization of long optical fibers. The determination of $OH^-$ levels in the sub-ppm range in small bulk samples, however, is prone to artifacts from $OH^-$ and adsorbed water on the sample surfaces.

Heating by background absorption (HBA) and differential luminescence thermometry (DLT) are two optical methods that are particularly useful for routine characterization and indirect measurement of impurities. In the HBA technique, the sample is irradiated by a high-power cw laser and the sample temperature is imaged by a high-resolution infrared camera. The laser is tuned to excite the problematic transition-metal ions while avoiding the active ion (if present). The observed heating is proportional to the background absorption coefficient, which is a measure of the total concentration of transition-metal ions and absorbing bulk defects. By using time-resolved thermal imaging it is also possible to distinguish internal sample heating from heating at the surfaces. In ZBLAN:$Yb^{3+}$, for example, HBA can be measured by using a multiwatt 1064-nm cw laser. While the 1064-nm excitation is close to the $Yb^{3+}$ absorption, it avoids excitation of the active ion since the $Yb^{3+}$ absorption is spectrally narrow. In contrast, the divalent transition-metal ions have spectrally broad lines, and their excitation at 1064 nm is representative of that at the cooling pump wavelength. Unwanted inclusions in the material, such as residual microscopic carbon particles or black precipitates of reduced zirconium, are also excited at this wavelength. Internal heating of more than ~ 1 mK/W in ZBLAN:$Yb^{3+}$ is indicative of excessive background absorption, and such samples typically do not exhibit useful laser cooling. The HBA technique is therefore an efficient tool for rapidly screening a large number of samples. In the DLT technique, the sample is excited at the cooling wavelength and the internal temperature is determined from the active ion luminescence spectrum. The typical setup is a single-pass configuration without mirrors deposited on the sample. The experiment uses a sequence of the pump laser being on for several seconds during which the sample heats or cools, followed by the pump laser off for several seconds during which the sample equilibrates back to the ambient temperature. Cycling the pump reduces the effects of long-term temperature drifts and decouples temperature changes inside the sample from those at the surfaces. Internal sample temperatures can be measured with ±10 mK accuracy in $Yb^{3+}$ laser cooling materials. As the ambient temperature is lowered, the cooling efficiency decreases, eventually reaching a critical temperature where laser cooling and residual heating are exactly balanced. This zero-crossing temperature is a measure of the purity and overall quality of the material.

## 2.4
## Glass Fabrication

### 2.4.1
### Glass Formation in ZrF$_4$ Systems

Heavy-metal fluoride glasses use $ZrF_4$ or $HfF_4$ as glass-formers to create a random three-dimensional network of edge- and corner-sharing $[ZrF_n]$ or $[HfF_n]$ polyhedra, where $n$ is typically 6 or 7. Pure $ZrF_4$ or $HfF_4$ do not exist in the vitreous form, and glass modifiers such as monovalent (NaF) or divalent ($BaF_2$) fluorides are needed to provide charge neutrality in the interstices of the network. Binary glasses are quite unstable and require fast rates of cooling, while ternary compositions such as $ZrF_4$-$BaF_2$-NaF readily form glasses. Such heavy-metal fluoride glasses were first discovered by Poulain et al., who reported glass formation for a composition of 50% $ZrF_4$-25% $BaF_2$-25% NaF (mol%) in the mid-1970s [111]. The family of ZBLAN glasses was developed during the following years by patient experimental trial and error. A relatively stable composition was found to be around 53% $ZrF_4$-20% $BaF_2$-4% $LaF_3$-3% $AlF_3$-20% NaF. This is often referred to as the "standard" ZBLAN composition. Trivalent rare-earth ions can substitute for $LaF_3$ in the ZBLAN host and enable a variety of optical applications. Figure 2.10 shows the $ZrF_4$-$BaF_2$-NaF phase diagram and the glass-forming region for ZBLAN. The $LaF_3$ and $AlF_3$ concentrations are fixed at 4 and 3%, respectively, to allow for the ternary plot. We calculate the phase diagram of Figure 2.10 using the devitrification theory developed by McNamara and Mair, who define the glass-stability parameter

$$G_s = (N_c - N)(X_c - X_g) , \qquad (2.3)$$

as a measure of glass stability [112]. $N$ and $X_g$ are parameters for nucleation and crystal growth, respectively, that depend on the glass composition and the bond enthalpies. $N_c = 8$ and $X_c = 1.9$ are critical nucleation and crystal growth parameters that are determined from an analysis of a large number of fluoride compositions. Stable glasses can form by slow cooling of a melt if $G_s > 0.4$. In Figure 2.10, the "standard" ZBLAN composition is clearly in the region of predicted maximum glass stability.

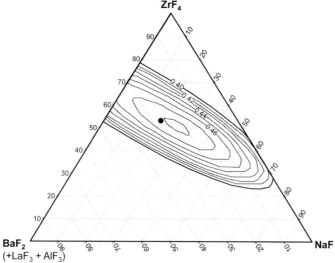

**Figure 2.10** Ternary phase diagram of ZBLAN glass showing our calculations of the contours of the glass stability parameter $G_s$ [112]. Glasses can be obtained by the slow cooling of a melt if $G_s > 0.4$. The *black circle* indicates the "standard" ZBLAN glass composition 53% $ZrF_4$-20% $BaF_2$-4% $LaF_3$-3% $AlF_3$-20% NaF. The concentrations of $LaF_3$ and $AlF_3$ were fixed at 4 and 3 mol %, respectively.

The preparation of ZBLAN glasses is complicated by their tendency to form black precipitates during melting in an inert atmosphere. The precipitates can occur as macroscopic black flakes or as a uniform gray tint. Robinson et al. first showed that the black phase is fluoride deficient with respect to the surrounding glass matrix [61], and Carter et al. subsequently used EPR measurements to identify considerable amounts of $Zr^{3+}$ in this phase [62]. The reduction of $Zr^{4+}$ to lower oxidation states occurs during glass melting and is accompanied by a loss of fluorine gas from the melt. The resulting black particles are unacceptable for optical applications, and optical refrigeration in particular, because they introduce excessive absorption that causes heating. A common solution to this problem is to prepare ZBLAN glasses under oxidizing conditions. Various approaches have been proposed over the years, including adding ammonium bifluoride ($NH_4 \cdot HF_2$) as a fluorinating agent or exposing the melt to $CCl_4$ or $O_2$ gaseous oxidizers. The latter approach is often referred to as *reactive atmosphere processing* and was first used for fluoride crystal growth. $CCl_4$ or $O_2$, however, introduce nonfluoride anions into the melt which can promote the crystallization of the glass. The use of $NH_4 \cdot HF_2$ was widespread in the early years of fluoride glass fabrication, but it has been abandoned altogether for the fabrication of high-purity glasses because of its many disadvantages [113]: (1) the fluorination reaction with $NH_4 \cdot HF_2$ produces gaseous water that can be incorporated into or react to $OH^-$ with the glass; (2) fluorination with $NH_4 \cdot HF_2$ is time-consuming and can take up to 15 h at 200–400 °C;

(3) $NH_4$ ions can be incorporated into the glass and act as vibrational impurities; (4) $NH_4 \cdot HF_2$ can be a source of transition-metal impurities; and (5) $NH_4 \cdot HF_2$ can lead to fuming of the melt and contamination of the furnace. $NF_3$ has been used successfully as an all-fluoride oxidizing gas. It has proven to be impractical, however, because it is very corrosive to furnace components at typical glass-melting temperatures. These problems can be avoided by using an in-situ fluoride oxidizer such as $InF_3$ or $SnF_4$ [63, 64]. Indium or tin are believed to oxidize zirconium according to

$$2ZrF_3 + InF_3 \rightarrow 2ZrF_4 + InF$$
$$2ZrF_3 + SnF_4 \rightarrow 2ZrF_4 + SnF_2.$$

As little as 0.25 mol % of $InF_3$ has been found to quantitatively suppress the reduction of zirconium in fluorozirconate melts under an inert atmosphere [64]. The resulting $In^+$ or $Sn^{2+}$ ions absorb at wavelengths around 300 nm, causing a small redshift of the ZBLAN band edge in the ultraviolet. However, these ions have no absorptions in the visible or infrared spectral range [64] and are benign in the context of laser cooling with rare-earth ions. The addition of in situ oxidizers to ZBLAN tends to decrease the glass stability (i.e. to increase the tendency of the glass to crystallize during quenching of the melt). Figure 2.11 illustrates the effect of adding $InF_3$ on the glass stability parameter $G_s$. The glass stability decreases as $InF_3$ is added, and the $G_s = 0.4$ threshold value for glass formation is reached at an $InF_3$ concentration of ~ 4 mol %. Rapid melt quenching is required beyond this concentration. $InF_3$ concentrations of up to 2.5 mol % are effective at preventing the formation of black precipitates while still allowing the casting of crystal-free glasses using standard melt quenching techniques.

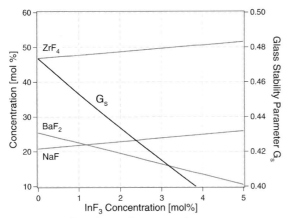

**Figure 2.11** Effect of $InF_3$ (the *in situ* oxidizer) concentration on the glass stability parameter, $G_s$, and on the optimum ZBLAN glass composition. $G_s$ was numerically maximized for each $InF_3$ concentration by varying the concentrations of $ZrF_4$, $BaF_2$, and NaF. The concentrations of $LaF_3$ and $AlF_3$ were fixed at 4 and 3 mol %, respectively.

## 2.4.2
## ZBLAN Glass Fabrication

The common method of fabricating bulk ZBLAN glass is to melt the appropriate metal-fluoride precursors in a furnace, cast the melt into a mold to form a glass by quenching, and anneal the glass to produce a stress-free sample. All of these steps have to occur under a strictly inert atmosphere. The use of an inert atmosphere prevents the oxidation of the melt via the reaction with atmospheric oxygen and water. This is critical, since oxide and $OH^-$ impurities can cause both non-radiative relaxation and crystallization of the glass. A second (although minor) effect is the dissolution of atmospheric gases in ZBLAN melts. The solubility for $N_2$, He, Ne, and Ar gas in a ZBLAN melt increases with decreasing melt temperature [114], and these gases typically remain dissolved upon melt quenching. $CO_2$ shows the opposite behavior [114] and can cause microbubbles during the quenching of the melt. Argon has one of the smallest solubilities of any gas in fluorozirconate melts ($< 10^{16}$ atoms/g) and is a common choice as an inert gas. The batch preparation, glass melting, casting, and annealing steps are usually performed inside a glove box that is actively controlled to sub-ppm water and oxygen levels. Starting materials are weighed, mixed, thoroughly ground in an agate mortar, and transferred into the melting crucible. The subsequent glass melting can be carried out in glassy carbon or noble metal crucibles such as platinum, gold, rhodium, and iridium. Lu et al. have documented submicrometer particles in ZBLAN glass and found them to be noble metal inclusions originating from the crucible [115]. These particles can cause unwanted scattering and absorption losses. Glassy carbon is therefore the preferred crucible material for melting fluoride glasses. It is inert under the conditions of interest and has excellent non-wetting properties.

Many factors have to be considered when designing the time–temperature profiles for ZBLAN glass melting and for casting and quenching of the melt. Some of these factors are conflicting, and a compromise has to be found by careful experimentation. The following sections describe some of the factors that must be considered.

### 2.4.2.1
**Melting of the Starting Materials**
With the exception of $ZrF_4$, all of the metal fluoride components of ZBLAN have melting temperatures of $> 996\,°C$. As the initial batch of mixed metal fluoride precursors is heated, $ZrF_4$ can lose fluorine and undergo unwanted reduction which, at this point, cannot be prevented by the in situ oxidizer (e.g. $InF_3$) since it has not yet melted and is therefore not available to reoxidize the zirconium. If there is no fluorination step with $NH_4 \cdot HF_2$ during melting, it is advantageous to use a high rate of heating when melting the initial batch of fluorides. One way to accomplish this is to insert the crucible directly into the furnace preheated to the melting temperature.

### 2.4.2.2
**Evaporative Losses**

$ZrF_4$ has a substantial vapor pressure (see Figure 2.9), and $ZrF_4$ evaporation can change the melt composition to the extent that the composition shifts outside the glass-forming range and causes crystallization during subsequent melt quenching. Melting in covered crucibles that have a small melt surface relative to the melt volume can reduce this effect. Melting above 850 °C for extended periods of time should be avoided unless evaporative $ZrF_4$ losses are accounted for in the composition of the initial batch of metal fluoride precursors. Routine monitoring of weight loss during glass melting is indispensible.

### 2.4.2.3
**Dissolution and Homogenization**

The glass batch has to be held in a molten state for a sufficiently long period of time for all starting materials to dissolve and for entrained gas to escape to the melt surface. This step is often referred to as *fining* of the melt and deserves particular attention, especially if the melt is not externally agitated. The typical melting temperature is well below the melting temperature of all metal fluoride precursors except that of $ZrF_4$. Even $ZrF_4$ does not have a sharply defined melting temperature as it develops a substantial vapor pressure at temperatures > 700 °C. The dynamics of melt formation, therefore, are complex and involve fluxing of the precursors in $ZrF_4$ vapor and subsequent dissolution of the materials in the $ZrF_4$-rich melt. Studies have shown that the extrinsic scattering loss of ZBLAN glass can be substantially reduced by increasing the melting time [116,117], indicating that residual particles from incomplete melting can be present in the final glass. ZBLAN:$Yb^{3+}$ glass that has not been melted for a sufficiently long period of time often shows visible $Yb^{3+}$ upconversion that is localized to microscopic particles. This upconversion likely occurs in residual $YbF_3$ crystals that are not fully dissolved during the melting process. The melting temperature should be high enough and the melting period long enough to ensure complete dissolution and homogenization of the constituents while avoiding excessive $ZrF_4$ evaporation.

### 2.4.2.4
**Optimum Rate of Cooling**

Compared to silica glasses, for example, heavy metal fluoride glasses such as ZBLAN are very susceptible to crystallization. A sufficiently high rate of cooling is necessary through the region of the crystallization and glass-transition temperatures to avoid the formation of microcrystallites. Melts have to be cooled at a rate of at least 0.2 K/s for even the most stable ZBLAN glass compositions to prevent the formation of crystallites. This limits the thickness of bulk samples to ~ 25 mm [118]. For larger samples it may not be possible to cool the melt quickly enough to prevent crystallization in the center of the shape. A high rate of cooling is obtained by lowering the temperature of the melt to near the liquidus temperature (~ 617 °C for ZBLAN [119]) before it is quenched during the casting process. A fast rate of cooling must however be balanced with the need to control internal stresses that develop during the solidification of the melt. An excessively fast cooling rate can lead to high internal stresses and cracking of the glass. Many authors report the need for casting into preheated molds to reduce the cooling rate and to avoid fracture of the glass. There is evidence that residual oxide particles play an important role not only in seeding the formation of fluoride crystals but also in seeding and propagating cracks during cooling. Melts with very low levels of oxide impurities, such as the ones prepared from HF gas fluorinated precursors (see Section 2.3.3.3), can be quenched at very high rates of cooling without inducing mechanical failure.

#### 2.4.2.5
**Viscosity for Casting**

The melt must have a sufficiently low viscosity at the temperature from which casting takes place so that the casting process can be completed before the melt solidifies. The intrinsically low viscosity of heavy metal fluoride melts above the liquidus temperature is helpful in this respect [120]. ZBLAN melts can usually be completely transferred from the melting crucible to the mold during the casting process. On the other hand, the low viscosity can cause the melt flow to become turbulent during casting and trap gas bubbles in the final glass. Another effect related to viscosity is the significant volume contraction that occurs when ZBLAN melts solidify. It can cause the formation of "vacuum bubbles" by cavitation. These bubbles form as a result of the external surface solidifying followed by volume contraction in the center of the sample. This becomes a significant problem only for relatively large sample volumes, and some reports suggest that cooling of the melt to below the liquidus temperature before casting may reduce this effect.

#### 2.4.2.6
**Typical Glass Fabrication Parameters**

A typical ZBLAN melting profile that balances these factors for gram-sized batches is as follows. A batch of metal fluoride precursors is prepared in a covered glassy carbon crucible and inserted into a 750 °C preheated furnace for 5 h. The temperature is lowered to 550 °C at a rate of 10 K/min, and the melt is cast into a platinum mold at room temperature. Glass produced by this method has considerable internal stresses, requiring subsequent annealing of the as-cast glass inside the casting mold. During the annealing process, the temperature is slowly raised to just below the glass transition temperature $T_g$. A typical annealing temperature is $T_g - 10$ K, which is at 250 °C for ZBLAN. The ion mobility is sufficiently high at this temperature to relieve internal stresses but still low enough to prevent crystallites from forming. The glass is held at the annealing temperature for 1 h and then slowly cooled to room temperature at a rate of 0.3 K/min. Poulain *et al.* have shown that annealing time appears to be less important than a slow rate of cooling for producing a stress-free glass [76].

In a final step, the desired sample shape is cut from the annealed ingot and an optical-quality polish is applied to all surfaces. Cutting can be done with diamond sectioning blades, but the cooling liquid has to be chosen carefully. Water-based liquids should be avoided because of the tendency of ZBLAN glass to rapidly hydrolyze in liquid water, and mixtures of glycerol and ethylene glycol are preferred. Optical-quality polishes can be realized readily using a sequence of commercial dry polishing pads of decreasing grit size. It has been found that dry polishing on aluminum oxide lapping films can embed microscopic particles of foreign material into the glass surface. This causes unacceptable surface heating when exposed to high laser pump power. Polishing with water-free alumina slurries of decreasing grit size significantly reduces this problem and produces surfaces that have both a high optical quality and negligible surface heating under high-power optical pumping.

### 2.4.3
### Fluoride, Chloride, and Sulfide Glass Fabrication

Many of the aspects of ZBLAN glass fabrication apply to the synthesis of other non-oxide glasses. Fluoroindate glasses (such as BIG), fluoroaluminate glasses (such as ABCYS) and fluorogallate glasses are also fabricated under a strictly inert atmosphere and follow annealing, cutting and polishing procedures that are similar to those used for ZBLAN. While fluoride glasses are fairly resistant to hydrolysis by gaseous water under ambient atmospheric conditions, fluorochloride glasses (such as CNBZn) and pure chloride glasses (such as $ZnCl_2$-based glasses) are significantly more hygroscopic and usually degrade within hours or days in ambient atmosphere. They require exclusive handling under an inert atmosphere at all stages of sample fabrication. Stable two-component sulfide glasses that use $Ga_2S_3$ as a glass former and $Na_2S$ as a glass modifier can be produced and may be attractive for laser-cooling applications. Melting of sulfide glasses typically occurs around 1100 °C in glassy carbon crucibles under an inert atmosphere.

## 2.5
## Halide Crystal Growth

The most common techniques for growing large bulk halide crystals are variations of the Czochralski or the Bridgman–Stockbarger methods [78]. In the Czochralski technique, a seed crystal is mounted on a holder and dipped into the melt. A crystal rod can be drawn from the melt by rotating and slowly translating the seed crystal upward in a process that requires exact control of temperature gradients and of the rates of pull and rotation. Growth rates can be several cm/h. The quality of the crystal can be sensitive to diameter fluctuations, and weight feedback systems are often incorporated to ensure a uniform diameter along the crystal rod. In the simpler Bridgman–Stockbarger method, a melt is slowly cooled from one end of the ampoule where a seed crystal is located. This is typically achieved by translating a two-zone furnace across the crystal growth ampoule. The crystal grows inside the ampoule as the melt solidifies in the temperature gradient, which can range from 10 to 100 K/cm [78]. Careful control of the temperature to within a few tenths of a degree is required and vibrations must be kept at a minimum. Both vertical and horizontal geometries are used, and growth rates are usually only a few mm/h. The quality of the crystal is often limited by the thermomechanical strain the crucible exerts on the crystal.

The laser cooling crystal $BaY_2F_8$ can be grown by either the Czochralski [33, 34, 121] or the Bridgman–Stockbarger [122] methods, with both yielding crystal rods several $cm^3$ in volume. The Czochralski growth of $BaY_2F_8$:$Yb^{3+}$ by Bigotta et al. was carried out in an argon and $CCl_4$ atmosphere to prevent the reduction of $Yb^{3+}$ to $Yb^{2+}$ [33, 34]. They rotated the crystal at a rate of 5 rpm and pulled at a rate of 0.5 mm/h from a melt at 995 °C, producing a crystal of high optical quality and free of cracks and microbubbles. Similarly, the laser cooling crystal $YLiF_4$ has been grown by either the Czochralski [123] or the Bridgman–Stockbarger [124] methods. As they did for $BaY_2F_8$:$Yb^{3+}$, Bigotta et al. used the Czochralski technique to grow $YLiF_4$:$Yb^{3+}$ at a rate of 1 mm/hr from a melt at 845 °C [35]. For both $BaY_2F_8$ and $YLiF_4$, the Bridgman–Stockbarger method tends to be simpler and can yield a more uniform distribution of the rare-earth dopants [125].

There are a few chloride crystals that show a surprisingly low degree of hygroscopicity and may be attractive for laser cooling with $Ce^{3+}$ and $Dy^{3+}$, for which a low-energy phonon material such as a chloride is required. These crystals include $KPb_2Cl_5$ [39] and the very toxic $Cs_3Tl_2Cl_9$ [126], both reported to be stable under ambient atmospheric conditions. These chlorides can be grown by the Bridgman–Stockbarger method, yielding crystals with $> 10 \, cm^3$ volumes. The growth of $KPb_2Cl_5$ from $KCl$ and $PbCl_2$ requires the removal of oxygen and water in the precursors to prevent the formation of lead hydroxychlorides and lead oxychlorides during heating. This can be accomplished by drying the precursors under vacuum at 130–150 °C for several days, growing an initial $KPb_2Cl_5$ crystal by repeated zone refinement in a horizontal furnace to segregate $KPb_2Cl_5$ from oxide and hydroxide impurities, and using the purified portion of the crystal for the subsequent Bridgman–Stockbarger growth of the rare-earth-doped $KPb_2Cl_5$ crystal [127, 128]. $KPb_2Cl_5$ has a high coefficient of thermal expansion and undergoes two phase transitions at 175 and 267 °C. As a result, internal stresses develop during cooling, making the crystals prone to cracking [129]. Careful control of the cooling rates was shown to partially mitigate this problem [129].

## 2.6
### Promising Future Materials

Doped ZBLAN glass has reached a high level of maturity as a laser-cooling material during the past decade. Advances in the purification of metal-fluoride precursors and glass fabrication methods will allow ZBLAN optical refrigerators to cool below 150 K in the near future. Despite this progress, ZBLAN may not be the optimum material for optical refrigerators in the long term because of two major drawbacks. First, ZBLAN glass has seven components when counting the rare-earth dopant and in situ oxidizer. This makes ZBLAN very laborious and expensive to prepare in high purity and creates many opportunities for contamination by impurities during the numerous processing steps. Materials with fewer components are highly desirable. Second, with a phonon cut-off energy of 580 cm$^{-1}$, ZBLAN is excellent for laser cooling with Yb$^{3+}$, increasingly marginal for the smaller energy gap of Tm$^{3+}$, and not workable for Ho$^{3+}$, Dy$^{3+}$ or Ce$^{3+}$ active ions (see Figure 2.3). Host materials with lower phonon cut-off energies are desirable for improved laser cooling with Tm$^{3+}$ and possible high-efficiency laser cooling with Ho$^{3+}$, Dy$^{3+}$ or Ce$^{3+}$. Several research avenues towards high-performance laser cooling materials are described in the following.

## 2.6.1
### Simplified Fluoride Glasses

While the "standard" ZBLAN composition yields a very stable fluorozirconate glass, fluoride glasses with fewer components may still be sufficiently stable to enable crystallite-free bulk samples for laser cooling applications. One approach is to simplify the ZBLAN composition by (1) omitting the 3 mol % AlF$_3$ component, (2) omitting the LaF$_3$ component and replacing it with the YbF$_3$ or TmF$_3$ active ion component, and (3) omitting NaF and only using BaF$_2$ as a modifier component. The result would be the four-component glass ZrF$_4$-YbF$_3$-InF$_3$-BaF$_2$. Using fixed 2% YbF$_3$ and 2.5% InF$_3$ concentrations, a maximum glass-stability parameter $G_s = 0.419$ is calculated for the optimized composition 63.3% ZrF$_4$-2% YbF$_3$-2.5% InF$_3$-32.2% BaF$_2$. This composition is expected to be a stable glass based on the $G_s > 0.4$ glass-forming criterion by McNamara et al. [112]. In fact, the glass stability parameter of this four-component glass is comparable to the $G_s = 0.424$ of the six-component ZBLAN composition optimized for 2.5% InF$_3$ (see Figure 2.11). Such simplified fluorozirconate glasses would greatly simplify the fabrication process and decrease the impurity level achievable in the final glass.

## 2.6.2
### Fluoride Crystals

Fluoride crystals are very attractive laser cooling hosts for Yb$^{3+}$ and Tm$^{3+}$ because of their simpler composition, lower phonon cut-off energy, and excellent thermal and mechanical properties in comparison to ZBLAN. These materials have already shown some of their potential (see Table 2.1), such as the recent cooling by 69 K in YLiF$_4$:Yb$^{3+}$ [36] or by 3.2 K in BaY$_2$F$_8$:Tm$^{3+}$ [38]. The use of high-purity LiF, YF$_3$, BaF$_2$, YbF$_3$ and TmF$_3$ precursors prepared by the methods described in Section 2.3, the implementation of ultraclean processing conditions during crystal growth, and the use of advanced polishing procedures are expected to significantly improve the laser cooling efficiency in these materials.

## 2.6.3
### Chloride and Bromide Crystals

Chlorides and bromides have attractive phonon properties but pose formidable challenges due to their extreme hygroscopicity. While some halides, such as LaBr$_3$:Ce$^{3+}$ gamma-ray scintillators, have been commercialized successfully, these challenging material properties are not amenable to high-volume device manufacturing. KPb$_2$Cl$_5$ is an exceptional crystal as it is only one of only a few nonhygroscopic halide materials. Yb$^{3+}$- and Er$^{3+}$-doped KPb$_2$Cl$_5$ crystals have already shown some laser-induced cooling (see Table 2.1), and here too the reduction of transition metal, hydroxide, and oxide impurities may enable significant advances. The very low phonon cut-off energy of only 200 cm$^{-1}$ in KPb$_2$Cl$_5$ makes this system a candidate for laser cooling with Ho$^{3+}$, Dy$^{3+}$, or Ce$^{3+}$ (see Figure 2.3). The high-power cw lasers needed to pump the Ce$^{3+}$ $^2F_{5/2} \rightarrow {}^2F_{7/2}$ transition around 5 µm pose an experimental problem. Quantum cascade lasers or OPOs based on ZnGeP$_2$ (ZGP) [130] might be possibilities. On the other hand, the 3-µm wavelength for pumping the Dy$^{3+}$ $^6H_{15/2} \rightarrow {}^6H_{13/2}$ transition is accessible by high-power cw OPOs. Furthermore, the 3-µm Dy$^{3+}$ transition has a very low spectral overlap with transition-metal ion absorptions (see Figure 2.6) and may be substantially less susceptible to quenching via such impurities. KPb$_2$Cl$_5$:Dy$^{3+}$ may be an attractive compromise between maximizing the cooling efficiency and choosing a practical pump laser.

## References

1 Pringsheim, P. (**1929**) Zwei Bemerkungen über den Unterschied von Lumineszenz- und Temperaturstrahlung, *Z. Phys.*, **57**, 739.

2 Landau, L. (**1946**) On the vibrations of the electronic plasma, *J. Phys.*, **10**, 25.

3 Kastler, A. (**1950**) Quelques suggestions concernant la production optique et la détection optique d'une inégalité de population des niveaux de quantification spatial des atoms. Application a l'expérience de Stern et Gerlach et a la resonance magnétique, *J. Phys. Radium*, **11**, 255.

4 Chu, S., Hollberg, L., Bjorkholm, J.E., Cable, A., Ashkin, A. (**1985**) Three-dimensional viscous confinement and cooling of atoms by resonance radiation pressure, *Phys. Rev. Lett.*, **55**, 48.

5 Kushida, T., Geusic, J.E. (**1968**) Optical refrigeration in Nd-doped yttrium aluminum garnet, *Phys. Rev. Lett.*, **21**, 1172.

6 Epstein, R.I., Buchwald, M.I., Edwards, B.C., Gosnell, T.R., Mungan, C.E. (**1995**) Observations of laser-induced fluorescent cooling of a solid, *Nature*, **377**, 500.

7 Hehlen, M.P., Epstein, R.I., Inoue, H. (**2007**) Model of laser cooling in the Yb$^{3+}$-doped fluorozirconate glass ZBLAN, *Phys. Rev. B*, **75**, 144302.

8 Tsoukala, V.G., Schroeder, J., Floudas, G.A., Thompson, D.A. (**1987**) Intrinsic Rayleigh scattering in fluoride glasses, *Mat. Sci. Forum*, **19–20** 637.

9 Thiede, J., Distel, J., Greenfield, S.R., Epstein, R.I. (**2005**) Cooling to 208 K by optical refrigeration, *Appl. Phys. Lett.*, **86**, 154107.

10 GOSNELL, T.R. (1999) Laser cooling of a solid by 65 K starting from room temperature, *Opt. Lett.*, **24**, 1041.
11 EDWARDS, B.C., ANDERSON, J.E., EPSTEIN, R.I., MILLS, G.L., MORD, A.J. (1999) Demonstration of a solid-state optical cooler: An approach to cryogenic refrigeration, *J. Appl. Phys.*, **86**, 6489.
12 LUO, X., EISAMAN, M.D., GOSNELL, T.R. (1998) Laser cooling of a solid by 21 K starting from room temperature, *Opt. Lett.*, **23**, 639.
13 MUNGAN, C.E., BUCHWALD, M.I., EDWARDS, B.C., EPSTEIN, R.I., GOSNELL, T.R. (1997) Laser cooling of a solid by 16 K starting from room temperature, *Phys. Rev. Lett.*, **78**, 1030.
14 RAYNER, A., FRIESE, M.E.J., TRUSCOTT, A.G., HECKENBERG, N.R., RUBINSZTEIN-DUNLOP, H. (2001) Laser cooling of a solid from ambient temperature, *J. Mod. Opt.*, **48**, 103.
15 MILLS, G.L., GLAISTER, D.S., GOOD, W.S., MORD, A.J. (2005) The performance of a laboratory optical refrigerator, *Cryocoolers*, **13**, 575.
16 HEEG, B., STONE, M.D., KHIZHNYAK, A., RUMBLES, G., MILLS, G., DEBARBER, P.A. (2004) Experimental demonstration of intracavity solid-state laser cooling of $Yb^{3+}$:$ZrF_4$-$BaF_2$-$LaF_3$-$AlF_3$-NaF glass, *Phys. Rev. A*, **70**, 21401.
17 GOOD, W.S., MILLS, G.L. (2004) Testing of samples for optical refrigeration, *Proc. SPIE*, **5554**, 153.
18 RAYNER, A., HIRSCH, M., HECKENBERG, N.R., RUBINSZTEIN-DUNLOP, H. (2001) Distributed laser refrigeration, *Appl. Opt.*, **40**, 5427.
19 SELETSKIY, D., HASSELBECK, M.P., SHEIK-BAHAE, M., EPSTEIN, R.I. (2007) Laser cooling using cavity enhanced pump absorption, *Proc. SPIE*, **6461**, 46104.
20 MURTAGH, M.T., SIGEL, G.H., FAJARDO, J.C., EDWARDS, B.C., EPSTEIN, R.I. (1999) Laser-induced fluorescent cooling of rare-earth-doped fluoride glasses, *J. Non-Cryst. Solids*, **253**, 50.
21 MUNGAN, C.E., BUCHWALD, M.I., EDWARDS, B.C., EPSTEIN, R.I., GOSNELL, T.R. (1997) Internal laser cooling of $Yb^{3+}$-doped glass measured between 100 and 300 K, *Appl. Phys. Lett.*, **71**, 1458.
22 FERNANDEZ, J., MENDIOROZ, A., GARCIA, A.J., BALDA, R., ADAM, J.L. (2000) Anti-Stokes laser-induced internal cooling of $Yb^{3+}$-doped glasses, *Phys. Rev. B*, **62**, 3213.
23 FERNANDEZ, J., MENDIOROZ, A., BALDA, R., VODA, M., AL-SALEH, M., GARCIA-ADEVA, A.J., ADAM, J.L., LUCAS, J. (2002) Origin of laser-induced internal cooling of $Yb^{3+}$-doped systems, *Proc. SPIE*, **4645**, 135.
24 GUIHEEN, J.V., HAINES, C.D., SIGEL, G.H., EPSTEIN, R.I., THIEDE, J., PATTERSON, W.M. (2006) $Yb^{3+}$ and $Tm^{3+}$-doped fluoroaluminate glasses for anti-Stokes cooling, *Phys. Chem. Glasses*, **47**, 167.
25 HOYT, C.W., PATTERSON, W., HASSELBECK, M.P., SHEIK-BAHAE, M., EPSTEIN, R.I., THIEDE, J., SELETSKIY, D. (2003) Laser cooling thulium-doped glass by 24 K from room temperature, *Trends Opt. Phot.*, 89, QThL4.
26 HOYT, C.W., HASSELBECK, M.P., SHEIK-BAHAE, M., EPSTEIN, R.I., GREENFIELD, S., THIEDE, J., DISTEL, J., VALENCIA, J. (2003) Advances in laser cooling of thulium-doped glass, *J. Opt. Soc. Am. B*, **20**, 1066.
27 HOYT, C.W., SHEIK-BAHAE, M., EPSTEIN, R.I., EDWARDS, B.C., ANDERSON, J.E. (2000) Observation of anti-Stokes fluorescence cooling in thulium-doped glass, *Phys. Rev. Lett.*, **85**, 3600.
28 FERNANDEZ, J., GARCIA-ADEVA, A.J., BALDA, R. (2006) Anti-Stokes laser cooling in bulk erbium-doped materials, *Phys. Rev. Lett.*, **97**, 033001.
29 GARCIA-ADEVA, A., BALDA, R., FERNANDEZ, J. (2007) Anti-Stokes laser cooling in erbium-doped low phonon materials, *Proc. SPIE*, **6461**, 646102.
30 BOWMAN, S.R., MUNGAN, C.E. (2000) New materials for optical cooling, *App. Phys. B*, B71, 807.
31 MUNGAN, C.E., BOWMAN, S.R., GOSNELL, T.R. (2000) Solid-state laser cooling of ytterbium-doped tungstate crystals, Int. Conf. on Lasers, 4–8 Dec. 2000, Albuquerque, NM, USA.
32 EPSTEIN, R.I., BROWN, J.J., EDWARDS, B.C., GIBBS, A. (2001) Measurements of optical refrigeration in

ytterbium-doped crystals, *J. Appl. Phys.*, 90, 4815.

33 BIGOTTA, S., PARISI, D., BONELLI, L., TONCELLI, A., TONELLI, M., DI LIETO, A. (2006) Spectroscopic and laser cooling results on $Yb^{3+}$-doped $BaY_2F_8$ single crystal, *J. Appl. Phys.*, 100, 13109.

34 BIGOTTA, S., PARISI, D., BONELLI, L., TONCELLI, A., DI LIETO, A., TONELLI, M. (2006) Laser cooling of $Yb^{3+}$-doped $BaY_2F_8$ single crystal, *Opt. Mat.*, 28, 1321.

35 BIGOTTA, S., DI LIETO, A., PARISI, D., TONCELLI, A., TONELLI, M. (2007) Single fluoride crystals as materials for laser cooling applications, *Proc. SPIE*, 6461, E1.

36 SELETSKIY, D., HASSELBECK, M.P., SHEIK-BAHAE, M., EPSTEIN, R.I., BIGOTTA, S., TONELLI, M. (2008) Cooling of Yb:YLF using cavity enhanced resonant absorption, *Proc. SPIE*, 6907, B9070.

37 PATTERSON, W., HASSELBECK, M.P., SHEIK-BAHAE, M., BIGOTTA, S., PARISI, D., TONCELLI, A., TONELLI, M., EPSTEIN, R.I., THIEDE, J. (2004) Observation of optical refrigeration in $Tm^{3+}$:$BaY_2F_8$, Lasers and Electro-Optics (CLEO), 16–21 May, San Francisco, CA, USA.

38 PATTERSON, W., BIGOTTA, S., SHEIK-BAHAE, M., PARISI, D., TONELLI, M., EPSTEIN, R.I. (2008) Anti-Stokes luminescence cooling of $Tm^{3+}$-doped $BaY_2F_8$, *Opt. Express*, 16, 1704.

39 MENDIOROZ, A., FERNANDEZ, J., VODA, M., AL-SALEH, M., BALDA, R., GARCIA-ADEVA, A.J. (2002) Anti-Stokes laser cooling in $Yb^{3+}$-doped $KPb_2Cl_5$ crystal, *Opt. Lett.*, 27, 1525.

40 WYBOURNE, B.G. (1965) Spectroscopic properties of rare earths, John Wiley & Sons, Inc., New York.

41 DIEKE, G.H., CROSSWHITE, H.M. (1963) The spectra of the doubly and triply ionized rare earths, *Appl. Opt.*, 2, 675.

42 A. JOULLIÉ, CHRISTOL, P., BARANOV, A.N., VICET, A. (2003) Mid-infrared 2–5 μm heterojunction laser diodes, *Topics Appl. Phys.*, 89, 1.

43 EBRAHIMZADEH, M. (2003) Mid-infrared ultrafast and continuous-wave optical parametric oscillators, *Topics Appl. Phys.*, 89, 179.

44 HÜFNER, S. (1978) Optical spectra of transparent rare-earth compounds, Academic, New York.

45 HOLLAS, J.M. (1987) Modern spectroscopy, John Wiley & Sons, Inc., New York.

46 RISEBERG, L.A., MOOS, H.W. (1968) Multiphonon orbit-lattice relaxation of excited states of rare-earth ions in crystals, *Phys. Rev.*, 174, 429.

47 VAN DIJK, J.M.F., SCHUURMANS, M.F.H. (1983) On the nonradiative and radiative decay rates and a modified exponential energy gap law for 4f–4f transitions in rare-earth ions, *J. Chem. Phys.*, 78, 5317.

48 LEI, G., ANDERSON, J.E., BUCHWALD, M.I., EDWARDS, B.C., EPSTEIN, R.I., MURTAGH, M.T., SIGEL, G.H. (1998) Spectroscopic evaluation of $Yb^{3+}$-doped glasses for optical refrigeration, *IEEE J. Quant. Electron.*, 34, 1839.

49 RICHARDSON, F.S., REID, M.F., DALLARA, J.J., SMITH, R.D. (1985) Energy levels of lanthanide ions in the cubic $Cs_2NaLnCl_6$ and $Cs_2NaYCl_6$:$Ln^{3+}$ (doped) systems, *J. Chem. Phys.*, 83, 3813.

50 GBOGI, E.O., CHUNG, K.H., MOYNIHAN, C.T., DREXHAGE, M.G. (1981) Surface and bulk OH infrared absorption in $ZrF_4$ and $HfF_4$ based glasses, *J. Am. Ceram. Soc.*, 64, C-51.

51 HUEBER, B., FRISCHAT, G.H., MALDENER, A., DERSCH, O., RAUCH, F. (1999) Initial corrosion stages of a heavy metal fluoride glass in water, *J. Non-Cryst. Solids*, 256&257 130.

52 RIZZATO, A.P., SANTILLI, C.V., PULCINELLI, S.H., MESSADDEQ, Y., CRAIEVICH, A.F., HAMMER, P. (2004) Study on the initial stages of water corrosion of fluorozirconate glasses, *J. Non-Cryst. Solids*, 348, 38.

53 SCHULTZ, P.C., VACHA, L.J.B., MOYNIHAN, C.T., HARBISON, B.B., CADIEN, K., MOSSADEGH, R. (1987) Hermetic coatings for bulk fluoride glasses and fibers, *Mat. Res. Forum*, 19–20, 343.

54 CHAUDHURI, R., SCHULTZ, P.C. (1987) Hermetic coatings on optical fibers, *Proc. SPIE*, 717, 27.

55 Hammer, P., Rizzato, A.P., Pulcinelli, S.H., Santilli, C.V. (2007) XPS study on water corrosion of fluorozirconate glasses and their protection by a layer of surface modified tin dioxide nanoparticles. *J. Electron Spectrosc.*, **156–158**, 128.

56 Tokiwa, H., Mimura, Y., Shinbori, O., Nakai, T. (1985) A core-clad composition for crystallization-free fluoride fibers, *J. Lightwave Tech.*, **LT-3**, 569.

57 Lima, S.M., Catunda, T., Lebullenger, R., Hernandes, A.C., Baesso, M.L., Bento, A.C., Miranda, L.C.M. (1999) Temperature dependence of thermo-optical properties of fluoride glasses determined by thermal lens spectrometry, *Phys. Rev. B*, **60**, 15173.

58 Aggarwal, R.L., Ripin, D.J., Ochoa, J.R., Fan, T.Y. (2005) Measurement of thermo-optic properties of $Y_3Al_5O_{12}$, $Lu_3Al_5O_{12}$, $YAlO_3$, $LiYF_4$, $LiLuF_4$, $BaY_2F_8$, $KGd(WO_4)_2$, and $KY(WO_4)_2$ laser crystals in the 80–300 K temperature range, J. Appl. Phys., **98** 103514.

59 Shurcliff, W.A. (1949) The trapping of fluorescent light produced within objects of high geometrical symmetry, *J. Opt. Soc. Am.*, **39**, 912.

60 Imbusch, G.F. (1978) Luminescence spectroscopy, (ed M.D. Lumb), Academic, London, 27.

61 Robinson, M., Pastor, R.C., Turk, R.R., Devor, D.P., Braunstein, M., Braunstein, R. (1980) Infrared-transparent glasses derived from the fluorides of zirconium, thorium, and barium, *Mat. Res. Bull*, **15**, 735.

62 Carter, S.F., France, P.W., Moore, M.W., Harris, E.A. (1987) Reduced species in glasses based on $ZrF_4$, *Phys. Chem. Glasses*, **28**, 22.

63 Hall, B.T., Andrews, L.J., Folweiler, R.C. (1989) Method for preparing fluoride glasses, U.S. Patent 4 946 490.

64 Andrews, L.J., Hall, B.T., Folweiler, R.C., Moynihan, C.T. (1988) In situ oxidation of heavy-metal fluoride glasses, *Halide Glasses V*, **32**, 43.

65 Poignant, H. (1987) Role of impurities in halide glasses, *NATO ASI, Series E*, **123**, 35.

66 Kaiser, P., Tynes, A.R., Astle, H.W., Pearson, A.D., French, W.G., Jaeger, R.E., Cherin, A.H. (1973) Spectral losses of unclad vitreous silica and soda-lime-silicate fibers, *J. Opt. Soc. Am.*, **63**, 1141.

67 France, P.W., Carter, S.F., Williams, J.R. (1984) Absorption in fluoride glass infrared fibers, *J. Am. Ceram. Soc.*, **67**, C243.

68 Poulain, M., Saad, M. (1984) Absorption loss due to complex anions in fluorozirconate glasses, *J. Lightwave. Tech.*, **LT-2**, 599.

69 Schultz, P.C., Vacha, L.J.B., Moynihan, C.T., Harbison, B.B., Cadien, K., Mossadegh, R. (1987) Hermetic coatings for bulk fluoride glasses and fibers, *Mat. Sci. Forum*, **19–20**, 343.

70 Jewell, J.M., Coon, J., Shelby, J.E. (1988) The extinction coefficient for $CO_2$ dissolved in a heavy metal fluoride glass, *Mat. Sci. Forum*, **32–33** 421.

71 France, P.W., Carter, S.F., Williams, J.R., Beales, K.J. (1984) OH-absorption in fluoride glass infrared fibers, *Electron. Lett.*, **20**, 607.

72 France, P.W., Carter, S.F., Parker, J.M. (1986) Oxidation states of 3d transition metals in $ZrF_4$ glasses, *Phys. Chem. Glasses*, **27**, 32.

73 Auzel, F. (1973) Materials and devices using double-pumped phosphors and energy transfer, *Proc. IEEE*, **61**, 758.

74 Goldner, P., Mortier, M. (2001) Effect of rare earth impurities on fluorescent cooling in ZBLAN glass, *J. Non-Cryst. Solids*, **284**, 249.

75 Mailhot, A.M., Elyamani, A., Riman, R.E. (1992) Reactive atmosphere synthesis of sol–gel heavy metal fluoride glass, *J. Mater. Res.*, **7**, 1534.

76 Poulain, M., Lebullenger, R., Saad, M. (1992) Synthesis of high purity fluorides by wet chemistry, *J. Non-Cryst. Solids*, **140**, 57.

77 Saad, M., Poulain, M. (1995) Fluoride glass synthesis by sol–gel process, *J. Non-Cryst. Solids*, **184**, 352.

78 Burkhalter, R., Dohnke, I., Hulliger, J. (2001) Growing of bulk crystals and structuring waveguides of fluoride

materials for laser applications, *Prog. Cryst. Growth Ch.*, **42**, 1–64.

79 Kwon, S.W., Kim, E.H., Ahn, B.G., Yoo, J.H., Ahn, H.G. (2002) Fluorination of metals and metal oxides by gas–solid reaction, *J. Ind. Eng. Chem.*, **8**, 477.

80 Krämer, K.W., Biner, D., Frei, G., Güdel, H.U., Hehlen, M.P., Lüthi, S.R. (2004) Hexagonal sodium yttrium fluoride based green and blue emitting upconversion phosphors, *Chem. Mater.*, **16**, 1244.

81 Bradley, D.C. (1989) Metal alkoxides as precursors for electronic and ceramic materials, *Chem. Rev.*, **89**, 1317.

82 Dejneka, M., Riman, R.E., Snitzer, E. (1993) Sol–gel synthesis of high-quality heavy-metal fluoride glasses, *J. Am. Ceram. Soc.*, **76**, 3147.

83 Turova, N.Y., Turevskaya, E.P., Kessler, V.G., Yanovskaya, M.I. (2002) The chemistry of metal alkoxides, Kluwer, Norwell.

84 Lafon, S., J.-Rajot, L., Alfaro, S.C. and Gaudichet, A. (2004) Quantification of iron oxides in desert aerosol, *Atmos. Environ.*, **38**, 1211.

85 Tolliver, D.L. (1988) Handbook of contamination control in microelectronics, William Andrew Inc., Norwich.

86 Hashimoto, Y., Amari, M., Komatsu, M., Fujiwara, K. (2005) Purification of trace amount of metal impurity from ultra pure water using membrane purifier/filter, *Solid State Phenom.*, **103–104**, 265.

87 Klein, P.H., Nordquist, P.E.R. Jr., Singer, A.H. (1985) Removal of cations from zirconium to permit its use in low loss fluoride optical fibers, *Opt. Eng.*, **24**, 516.

88 Sommers, J.A., Perkins, V.Q. (1988) Research and development for commercial production of high-purity $ZrF_4$ and $HfF_4$, *Mat. Sci. Forum*, **32–33**, 629.

89 Malissa, H., Schöffmann, E. (1955) Über die Verwendung von substituierten Dithiocarbamaten in der Mikroanalyse, *Mikrochim. Acta*, **1**, 187.

90 Kinrade, J.D., Van Loon, J.C. (1974) Solvent extraction for use with flame atomic absorption spectrometry, *Anal. Chem.*, **46**, 1894.

91 Brooks, R.R., Hoashi, M., Wilson, S.M., Zhang, R.Q. (1989) Extraction into methyl isobutyl ketone of metal complexes with ammonium pyrrolidine dithiocarbamate formed in strongly acidic media, *Anal. Chim. Acta*, **217**, 165.

92 Bertrand, D., Guery, J., Jacoboni, C. (1993) Fe, Co, Ni, Cu trace metal analysis in ZBLAN fluoride glasses, *J. Non-Cryst. Solids*, **161** 32.

93 Newman, P.J., Voelkel, A.T., MacFarlane, D.R. (1995) Analysis of Fe, Cu, Ni, and Co in fluoride glasses and their precursors, *J. Non-Cryst. Solids*, **184**, 324.

94 Kobayashi, K. (1988) Purification of raw materials for fluoride glass fibers by solvent extraction, *Mat. Sci. Forum*, **32–33**, 75.

95 Arnac, M., Verboom, G. (1974) Solubility product constants of some divalent metal ions with ammonium pyrrolidine dithiocarbamate, *Anal. Chem.*, **46**, 2059.

96 Ling, Z., Chengshan, Z., Gaoxian, D., Kangkang, W. (1992) $ZrOCl_2$ for fluoride glass preparation, *J. Non-Cryst. Solids*, **140**, 331.

97 Rodriguez, A.M., Martinez, J.A., Caracoche, M.C., Rivas, P.C., Lopez Garcia, A.R., Spinelli, S. (1984) Time-differential perturbed angular correlations investigation of the $(NH_4)_2ZrF_6$ thermal decomposition, *J. Chem. Phys.*, **82**, 1271.

98 Sense, K.A., Alexander, C.A., Bowman, R.E., Filbert, R.B. Jr. (1957) Vapor pressure and derived information of the sodium fluoride-zirconium fluoride system, *J. Phys. Chem.*, **61**, 337.

99 Kent, R.A., Zmbov, K.F., A.S. Kana'an, Besenbruch, G., McDonald, J.D., Margrave, J.L. (1966) Mass spectrometric studies at high temperatures. The sublimation pressures of scandium, yttrium, and lanthanum fluorides, *J. Inorg. Nucl. Chem.*, **28**, 1419.

100 Hart, P.E., Searcy, A.W. (1966) The vapor pressure, the evaporation coefficient, and the heat of sublimation of barium fluoride, *J. Phys. Chem.*, **70**, 2763.

101. Yaws, C.L. (2007) Handbook of vapor pressure, Gulf Publishing Company, Houston.
102. Brunetti, B., Villani, A.R., Piacente, V., Scardala, P. (2000) Vaporization studies of lanthanum trichloride, tribromide, and triiodide, *J. Chem. Eng. Data*, **45**, 231.
103. Bardi, G., Brunetti, B., Piacente, V. (1996) Vapor pressure and standard enthalpies of sublimation of iron difluoride, iron dichloride, and iron dibromide, *J. Chem. Eng. Data*, **41**, 14.
104. Brunetti, B., Piacente, V. (1996) Torsion and Knudsen measurements of cobalt and nickel difluorides and their standard sublimation enthalpies, *J. Alloys Comp.*, **236**, 63.
105. Brunetti, B., Piacente, V., Scardala, P. (2008) Vapor pressures and sublimation enthalpies of copper difluoride and silver (I,II) fluorides by the torsion-effusion method, *J. Chem. Eng. Data*, **53**, 687.
106. MacFarlane, D.R., Newman, P.J., Voelkel, A. (2002) Methods of purification of zirconium tetrafluoride for fluorozirconate glass, *J. Am. Ceram. Soc.*, **85**, 1610.
107. Bao, S., Newman, P.J., Voelkel, A., Zhou, Z., MacFarlane, D.R. (1995) Electrochemical purification and GFAAS analysis of heavy metal fluoride glass, *J. Non-Crys. Solids*, **184**, 194.
108. Fajardo, J.C., Sigel, G.H. Jr., Edwards, B.C., Epstein, R.I., Gosnell, T.R., Mungan, C.E. (1997) Electrochemical purification of heavy metal fluoride glasses for laser-induced fluorescent cooling applications, *J. Non-Cryst. Solids*, 213&214, 95.
109. Newman, P.J., Voelkel, A.T., MacFarlane, D.R. (1995) Analysis of Fe, Cu, Ni, and Co in fluoride glasses and their precursors, *J. Non-Cryst. Solids*, **184**, 324.
110. Bertrand, D., Guery, J., Jacoboni, C. (1993) Fe, Co, Ni, Cu trace metal analysis in ZBLAN fluoride glasses, *J. Non-Cryst. Solids*, **161**, 32.
111. Poulain, M., Poulain, M., Lucas, J., Brun, P. (1975) Verred fluores au tetrafluorure de zirconium. Proproetes optiques d'un verre dope au $Nd^{3+}$, *Mat. Res. Bull*, **10**, 243.
112. McNamara, P., Mair, R.H. (2005) Devitrification theory and glass-forming phase diagrams of fluoride compositions, *Proc. SPIE*, **5650**, 123.
113. Poulain, M., Soufiane, A., Messaddeq, Y., Aegerter, M.A. (1992) Fluoride glasses: Synthesis and properties, *Braz. J. Phys.*, **22**, 205.
114. Jewell, J.M., Shelby, J.E. (1987) Gas solubility in heavy metal fluoride melt, *Mat. Sci. Forum*, **19–20**, 287.
115. Lu, G., Aggarwal, I. (1988) Noble metals as a source of continuous scattering in fluoride glasses, *J. Am. Ceram. Soc.*, **71**, C156.
116. Nakai, T., Mimura, Y., Tokiwa, H. (1985) Shinbori, O., Origin of excess scattering in fluoride glasses, *J. Lightwave Tech.*, **LT-3**, 565.
117. Kanamori, T., Hattori, H., Sakaguchi, S., Ohishi, Y. (1986) Study on the nature of extrinsic scattering centers in fluoride glass optical fibers, *Jpn. J. Appl. Phys.*, **25**, L203.
118. Drexhage, M.G., Cook, L.M., Margraf, T., Chaudhuri, R., Schultz, P.C. (1988) Multikilogramm fluoride glass synthesis, *Halide Glasses V*, **32**, 9.
119. Parker, J.M. (1989) Fluoride glasses, *Annu. Rev. Mater. Sci.*, **19**, 21.
120. Yano, T., Inoue, S., Kawazoe, H., Yamane, M. (1988) Viscosity of the melt of $ZrF_4$-$BaF_2$-$LaF_3$-$AlF_3$-NaF (or LiF) systems, *Mat. Sci. Forum*, **32–33**, 583.
121. Guilbert, L.H., Gesland, J.Y., Bulou, A., Retoux, R. (1993) Structure and Raman spectroscopy of Czochralski-grown barium yttrium and barium ytterbium fluoride crystals, *Mat. Res. Bull.*, **28**, 923.
122. Auzel, F., Kaminskii, A., Meichenin, D. (1992) Diode and dye pumped cw laser in $BaY_2F_8$:$Er^{3+}$ at 2.7 μm, *Phys. Stat. Sol. A*, **131**, K63.
123. Ranieri, I.M., Baldochi, S.L., Santo, A.M.E., Gomes, L., Currol, L.C., Tarelho, L.V.G., de Rossi, W., Berretta, J.R., Costa, F.E., Nogueira, G.E.C., Wetter, N.U., Zezell, D.M., Vieira, N.D. Jr, Morato, S.P. (1996) Growth of $YLiF_4$ crystals doped with holmium,

erbium, and thulium, *J. Cryst. Growth*, **166**, 423.

**124** Misiak, L.E. (1997) On grystal growth of LiYF$_4$, *Proc. SPIE*, **3178**, 48.

**125** Rogin, P., Hulliger, J. (1997) Growth of LiF$_4$ by the seeded vertical gradient freezing technique, *J. Cryst. Growth*, **172**, 200.

**126** Kamber, L., Egger, P., Trusch, B., Giovanoli, R., Hulliger, J. (1998) High temperature phase segregation of a new host for Er$^{3+}$ upconversion: Cs$_3$Tl$_2$Cl$_9$, *J. Mater. Chem.*, **8**, 1259.

**127** Roy, U.N., Cui, Y., Guo, M., Groza, M., Burger, A., Wagner, G.J., Carrig, T.J., Payne, S.A. (2003) Growth and characterization of Er-doped KPb$_2$Cl$_5$ as laser host crystal, *J. Cryst. Growth*, **258**, 331.

**128** Amedzake, P., Brown, E., Hömmerich, U., Trivedi, S.B., Zavada, J.M. (2008) Crystal growth and spectroscopic characterization of Pr-doped KPb$_2$Cl$_5$ for mid-infrared laser applications, *Mat. Sci. Eng. B*, **146**, 110.

**129** Condon, N.J., S. O'Connor, Bowman, S.R. (2006) Growth and characterization of single-crystal Er$^{3+}$:KPb$_2$Cl$_5$ as a mid-infrared laser material, *J. Cryst. Growth*, **291**, 472.

**130** Godard, A. (2007) Infrared (2–12 µm) solid-state laser sources: a review, *C.R. Physique*, **8**, 1100.

# 3
# Laser Cooling in Fluoride Single Crystals
*Stefano Bigotta and Mauro Tonelli*

## 3.1
## Introduction

The first experimental evidence of the laser cooling of solids was provided by Epstein *et al.* [1] more than a decade ago, when the authors succeeded in cooling an Yb-doped fluorozirconate glass (ZBLANP) by anti-Stokes luminescence. Since then, progress has been made, and the temperature drops observed have increased from tenths of a Kelvin up to 88 K [2].

These results have encouraged several groups to search for new materials that could improve on the results obtained from Yb-doped ZBLANP. Indeed, up to now the most commonly investigated materials for laser cooling have been the fluorozirconate glasses ZBLAN and the very similar ZBLANP. Although a few other Yb-doped hosts have been found, several papers have reported on how the search has been largely unfruitful (see for example [3]). Indeed, even though a number of candidates for optical refrigeration have been identified, available samples often lack sufficient purity to show net cooling. Up to now, bulk cooling in Yb-doped crystals has been measured only in $Yb^{3+}$:YAG and $Yb^{3+}$:$Y_2Si_2O_5$ [4]. The thermal and mechanical properties of these crystals may be advantageous for practical applications, such as in the optical cryocooler [5], but the authors report that the cooling efficiency was not as high as in Yb-doped ZBLAN glasses, and the temperature drops observed in a single-pass configuration were 0.36 and 1 K for $Yb^{3+}$:YAG and $Yb^{3+}$:$Y_2Si_2O_5$, respectively.

All of the abovementioned experiments employed $Yb^{3+}$ as coolant, but this does not mean that $Yb^{3+}$ is the only possible choice. Indeed, the first rare-earth ion ever proposed was the $Gd^{3+}$ ion, as suggested by Yatsiv [6]; furthermore, the use of a ionic dopant with a smaller gap between the ground state and the excited-state manifold should lead to a higher cooling efficiency. In fact, just after the discovery of the laser cooling effect in solids, Edwards *et al.* [7] suggested the use of $Tm^{3+}$:ZBLANP or $Dy^{3+}$:$LaBr_3$ as cooling materials, since the smaller energy gaps between ground and first excited states of these materials should, in principle, lead to higher cooling efficiencies (see Section 3.5.1). However, the cooling of a $Tm^{3+}$-doped glass has only been demonstrated relatively recently, due to the novel availability of power-

*Optical Refrigeration. Science and Applications of Laser Cooling of Solids.*
Edited by Richard Epstein and Mansoor Sheik-Bahae
Copyright © 2009 WILEY-VCH Verlag GmbH & Co. KGaA, Weinheim
ISBN: 978-3-527-40876-4

ful laser sources in the mid-infrared. Hoyt et al. [8] observed a temperature drop of 1.2 K from room temperature for a single pass of the pump beam in a Tm:ZBLANP sample. In a more refined experiment [9], the authors allowed the pumping radiation to pass through the sample multiple times by placing it in a nonresonant cavity. Using this experimental setup, the authors obtained a cooling of 19 K from room temperature.

There is another class of materials that has been investigated: the chloride crystals. Local cooling has already been observed in $Yb^{3+}$:$KPb_2Cl_5$ [10], and a temperature drop of 0.5 K was recently measured in a 0.5% $Er^{3+}$:$KPb_2Cl_5$ sample [11].

In this chapter, we propose a different class of materials that can be used effectively in laser cooling application: single-crystal fluorides. The requirements that a candidate for a laser cooling host must fulfill are essentially high purity and high quantum efficiency. Indeed, the average amount of heat extracted per absorbed photon is very small and roughly equal to the thermal energy $kT$. Absorption by impurities and nonradiative transitions, in contrast, can transfer a large amount of energy to the host matrix, leading to sample heating. In addition to suitable physical properties, the thermal and mechanical properties of the host must be suited to practical applications.

High quantum efficiencies may be obtained using materials with low-energy phonons and using effective growth methods that avoid the inclusion of impurities or defects. For the same reason, a good material should not be hygroscopic, in order to avoid the presence of water molecules, which can significantly reduce the quantum efficiency. Furthermore, the candidate material should have good mechanical hardness: this not only allows the realization of a good surface finish, ensuring that heat is not generated at the interfaces, but it is also required in order to process the material. Finally, good thermal conductivity is needed to extract heat from the load when it is used in an optical cryocooler.

A further aid to choosing the right material is the observation that the anti-Stokes optical cycle is, in principle, a laser cycle run in reverse. This may suggest that a good laser host material that fulfills the above requirements could also be an efficient host for optical refrigerators. However, it is clear that the skills, the knowledge, and the technology employed in crystal growth are of fundamental importance when attempting to obtain new and fruitful materials.

The fluoride single crystals fulfill all of these requirements. They have long been shown to be a good laser host material [12, 13]. In addition, fluoride single crystals have low-energy phonons, are not hygroscopic, and show good mechanical and thermal properties, such as the high thermal conductivity required for laser cooling applications. Furthermore, the substantial and successful growth of very high quality crystals has enabled new and interesting results with rare-earth-doped fluoride crystals to be obtained [14–17]. Some of these results can easily be ascribed to the very high optical quality of the crystals grown, and to their lack of impurities and imperfections.

In this chapter, we report on recent progress in the field obtained with single crystals doped with Yb$^{3+}$ and Tm$^{3+}$, which were grown in a Czochralski furnace at the Physics Department of Pisa University.

Starting from their mechanical and physical properties, and using spectroscopic data we collected from our samples, we demonstrate that fluoride single crystals fulfill the requirements of laser cooling materials. The theoretical and experimental cooling efficiencies of these materials are then evaluated and compared with those of ZBLAN.

## 3.2
## Physical Properties

Two different fluoride crystals were investigated during the course of our work: BaY$_2$F$_8$ (BYF) and LiYF$_4$ (YLF).

BYF has a monoclinic crystalline structure with the $C_{2h}^3$ (C2/m) symmetry group [18]. The reticular constants are $a = 0.6935$ nm, $b = 1.0457$ nm, $c = 0.4243$ nm, with an angle $\gamma$ between the a-axis and the c-axis of 99.7°. The primitive cell contains two BYF molecules.

YLF has the scheelite structure with a tetragonal system. It crystallizes in the $I4_1/a$ ($C^6$ 4h) space group, and the lattice parameters are $a = 0.5155$ nm, $b = 1.068$ nm [19].

The most striking features of the fluoride single crystals with respect to ZBLAN are their very low phonon energies (~ 400 cm$^{-1}$ for BYF [20] and ~ 450 cm$^{-1}$ for YLF [21]), which makes them suitable for laser applications in the near-IR region, and their low thermal lensing. The mechanical and physical properties of the fluorozirconate glass ZBLAN are reported in Table 3.1 and compared with those of the crystals studied here (BYF and YLF), a well-known oxide crystal (YAG), and the chloride crystal KPb$_2$Cl$_5$. This table illustrates that fluoride crystals offer several important advantages over ZBLAN that make them attractive for laser cooling applications. Indeed, their lower phonon energy reduces the nonradiative decay rate of the quenching centers and indirectly leads to higher emission efficiency; their transparency at longer infrared wavelengths makes them less susceptible to deleterious radiative heat loading; their thermal conductivity is an order of magnitude higher; their slightly higher material hardness allows for better polishing and direct deposition of optical coatings; finally, their nonhygroscopic nature guarantees longer durability and higher efficiency.

Fluoride crystals show different mechanical properties compared to YAG. Specifically, YAG crystals show better thermal conductivity and higher material hardness. However, the optical properties of fluoride crystals can largely compensate for these advantages of YAG. Indeed, we believe that their lower phonon energy, broader transparency range and lower refractive index can make fluoride crystals rather than YAG good competitors to ZBLAN. On the other hand, if we compare fluoride single crystals with KPb$_2$Cl$_5$, we note that the chloride shows lower phonon energy

**Table 3.1** Physical properties of ZBLAN, YAG, BYF, YLF and KPb$_2$Cl$_5$ matrices (after [4, 21–25] and H. P. Jenssen, personal communication) (n.a. = not available).

| | ZBLAN | YAG | BaY$_2$F$_8$ | LiYF$_4$ | KPb$_2$Cl$_5$ |
|---|---|---|---|---|---|
| Crystal structure | Amorphous | Cubic | Monoclinic | Tetragonal | Monoclinic |
| Density (g/cm$^{-3}$) | 4.33 | 4.53 | 4.97 | 3.99 | 4.63 |
| Thermal conductivity (W/mK) | 0.9 | 13 | 6 | 6.3 | 4 |
| Transparency range (µm) | 0.2–7 | 0.25–5 | 0.125–12 | 0.18–6.7 | 0.2–20 |
| Refractive index at 1 µm | 1.5 | 1.83 | 1.51 (ave.) | 1.4 (ave.) | 2.01 (ave) |
| Microhardness (Knoop) | 225 | 1215 | 235–350 | 300 | 100 |
| Effective phonon energy (cm$^{-1}$) | 506 | 630 | 400 | 450 | 200 |
| Hygroscopic? | Yes | No | No | No | No |
| Specific heat (J g$^{-1}$ K$^{-1}$) | 0.596 | 0.59 | 0.5 | 0.79 | n.a. |

and a wider transparency range. However, its poor thermomechanical properties, along with its high refractive index, make it less useful for practical applications.

## 3.3
## Experimental

### 3.3.1
### Growth Apparatus

All of the crystals used for laser cooling were grown at the Physics Department of the University of Pisa in a homemade computer-controlled Czochralski furnace with resistive heating, as shown in Figure 3.1. As already mentioned, the requirements for the materials used in laser cooling applications are high purity, high optical quality and the ability to achieve a good optical polish. These characteristics should guarantee the absence of spurious absorbing and scattering centers and other optical defects that may lead to local heating in the crystal, which eventually spoils the cooling process. For this reason, special care was taken throughout the growth process, starting from the choice of the starting materials. To avoid OH$^-$ radical contamination, which significantly deteriorates the optical characteristics, the powders used as raw material were fluorinated and purified at AC Materials (Tampa, FL, USA), and a purity of 99.999% was obtained. Then all of the materials were stored and manipulated in special environments.

In order to stop OH$^-$ and other pollutants from contaminating the powders during the growth process, the growth chamber is equipped with a vacuum apparatus with an ultimate pressure limit of few $10^{-5}$ Pa. After efficient removal of air from the furnace, growth is usually performed in a inert atmosphere of high-purity argon and other gases, if needed, depending on the host and/or dopant.

The growth chamber for fluoride single crystals contains a graphite resistance heater, a crucible made of vitreous carbon, and some thermal shields that prevent

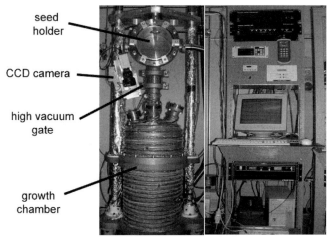

**Figure 3.1** The homemade Czochralski furnace used to grow the fluoride single crystals studied in this work; the furnace is shown on the *left* and the PC-assisted control system on the *right*.

the heat from dissipating far from the crucible. The highest temperature that can be achieved with this system is about 1200 °C. The main difference between our furnace and commercially available ones is the presence of a second chamber connected to the first by a high-vacuum gate, which allows to access to the interior of the furnace without the need to cool to room temperature and degrade the internal atmosphere. This unusual feature allows the melt to be carefully cleaned and impurity particles to be removed more efficiently than in conventional Czochralski systems and without wasting growth material.

The other two main parts of the furnace are the pulling system and the diameter control apparatus. The pulling system consists of a stainless steel rod fixed to a trolley moving along two stainless steel runners and equipped with two step motors: one for vertical motion and one for rotation. In order to ensure that a single crystal is grown and to reduce the internal stresses, the seed employed is an oriented single crystal too.

The diameter of the growing crystal is controlled using an optical system that is placed outside the furnace. It consists of a red diode laser that is directed onto the solid–liquid interface meniscus. As the diameter increases, the meniscus moves towards the crucible wall, and so does the reflection of the beam. A CCD camera acquires the bright point of the laser reflected by the meniscus, and a suitable algorithm transforms the position of the reflection into diameter information. Control is achieved via a PID feedback algorithm.

The right choice of values for the last two parameters (the pulling and rotation rates, along with the parameters of the PID algorithm) is of crucial importance to a successful growth. The use of wrong values can result in the introduction of crystal defects, imperfections, microbubbles and stresses in general. A clear correlation between defects and growth parameters is demonstrated, for instance, in [26]. In our case, crystal growth was carried out in a high-purity argon atmosphere in order to avoid melt contamination. For Yb-doped crystals, a suitable amount of $CF_4$ gas was also added in order to prevent the reduction of the $Yb^{3+}$ ions to $Yb^{2+}$. $Yb^{2+}$ ions introduce new channels of energy transfer that severely compromise the optical quality of the crystals, the quantum efficiency and thus the efficiency of the cooling process. The temperature of the melt was around 995 °C for BYF and 845 °C for YLF. During the growth, the rotation rate of the sample was 5 rpm and the pulling rates were 0.5 and 1 mm/h for the BYF and YLF, respectively. The crystals were grown using $LiF$-$YF_3$ and $BaY_2F_8$-$BaF_2$ powders as raw material for the crystal, and the required doping density was achieved by adding an appropriate amount of $YbF_3$ or $TmF_3$ powders. 2.5% Yb:BYF, 5% Yb:YLF and 1.2% Tm:BYF were grown using this apparatus.

The crystals were of a high optical quality; they were free of cracks and microbubbles. The Laue X-ray technique was used to check that the samples were single crystals, by identifying the crystallographic axes of the crystal and cutting oriented samples. For laser cooling experiments, small parallelepipeds were cut and all six faces were polished to a high optical quality.

### 3.3.2
**Spectroscopic Setup**

Room-temperature absorption spectra were acquired using a Cary 500 spectrometer. The resolution was better than 1 nm. Room-temperature fluorescence spectra were measured by exciting the sample using a suitable laser diode (centered at 970 and 790 nm for Yb and Tm samples, respectively). The luminescence was mechanically chopped and focused by a lens of 75 mm focal length onto the input slit of a 1/3 m monochromator equipped with a 600 groove/mm grating. The fluorescence was observed perpendicular to the exciting beam to reduce spurious pump scattering. The signal was detected by a cooled InSb detector and processed by a lock-in amplifier. The resolutions of the fluorescence spectra were 0.1 and 0.5 nm at around 1 and 2 µm, respectively. All of the polarized spectra were normalized relative to the absorbed power, and a special mounting was used to ensure that the same experimental conditions were employed for all orientations of the sample. The spectra were corrected for the spectral response of the system using a blackbody source at 3000 K.

For the fluorescence lifetime measurements, the samples were excited by a pulsed tunable Ti:sapphire laser or by a mechanically chopped 970 nm laser diode for Tm and Yb samples, respectively. In order to observe a uniformly pumped volume and to reduce the artifacts of radiation trapping, we collected the fluorescence from a short portion ($\approx$ 1 mm thick) of the sample; furthermore, the power incident on the crystal was reduced as much as possible by means of an attenuator in order to minimize nonlinear effects. The signal was detected by the same experimental apparatus as described above, and the signal was sent from a fast amplifier to a digital oscilloscope connected to a computer. The response time of the system was about 1 µs.

### 3.3.3
### Cooling Setup

The apparatus used in the single-pass laser cooling experiments, shown in Figure 3.2, consists of a laser emitting at an appropriate wavelength, whose output is focused onto the crystal. For the Yb-doped samples, the pump source consists of of two diode lasers mounted with the junctions perpendicular to each oth-

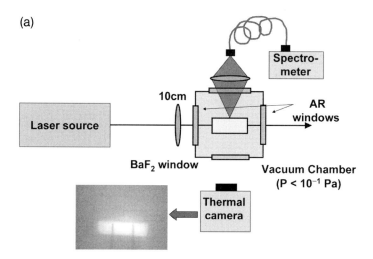

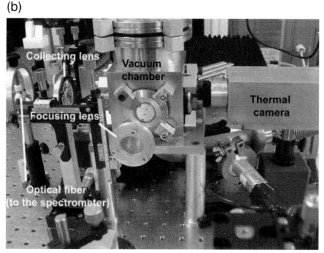

**Figure 3.2** Experimental setup used in laser cooling experiments: scheme (a) and picture (b). The sample is kept in a vacuum chamber in order to minimize parasitic heat transfer from the environment. Two different methods were used to monitor the temperature of the sample: those based on the use of a thermal camera and a fiber pigtailed spectrophotometer.

er. The output wavelength can be tuned slightly by changing the temperature of the diode junction, and is centered at approximately 1030 nm, with a bandwidth of ~ 2 nm used for both lasers. A polarizing beam splitter and a 10 cm focal length lens are used to recombine and focus the beams onto the crystal. The maximum power available just before the focusing lens is ~ 3 W. For measurements performed on the Tm-doped sample at the the Physics and Astronomy Department of the University of New Mexico, the pump is a homemade OPO, synchronously pumped by 80 ps (FWHM) mode-locked pulses from a $Nd^{3+}$:YAG laser at a wavelength of 1.064 µm with a repetition rate of 76 MHz, delivering up to 20 W. The OPO output is tunable (1.7 – 2.1 µm), and has a power output of ~ 4 – 6 W. The crystal is placed in a small vacuum chamber ($10 \times 10 \times 10\,cm^3$) and suspended on two crossed microscope coverslips in order to reduce parasitic heat loading from the surroundings, which occurs mainly through conduction and convection mechanisms. For this reason, the vacuum inside the chamber was kept at pressures of less than $10^{-1}$ Pa through the use of a cryogenic or a turbomolecular pump. Anti-reflection coated BK7 windows were employed in order to minimize power losses. To keep the heat load as low as possible, the temperature measurements were performed using noncontact thermometry. Two different methods were employed. The first one utilizes a long-wavelength infrared (LWIR) Raytheon Control IR 2500AS video camera placed outside the vacuum chamber. In order to allow LWIR radiation to reach the camera, the vacuum chamber was equipped with windows that are transparent at such wavelengths ($BaF_2$ or NaCl). The output of the thermal camera is a false-color video, where hot objects show up as dark regions and cold objects show up as light regions. The nominal resolution of the camera is 0.2 K. The data were then transferred to a PC by a videograbber board and stored in a 8-bit pixel image file. In order to obtain a proportionality relation between the pixel values and the temperature drop, the crystals were placed in a He closed-cycle cryostat, where the temperature of the sample was changed and monitored by a LakeShore 330 temperature controller. In the meantime, the thermal camera was used to obtain the images, and the acquisition conditions were chosen to be as close as possible to the experimental ones. From the calibration we noted that the thermal camera response is linear in the range $\Delta T \sim \pm 10$ K around room temperature. For temperatures that are less than 10 K below room temperature, the thermal camera saturates and cannot be used to determine the sample temperature. For larger temperature drops we used a different approach that deduces the temperature of the sample by comparing the intensity ratio of the fluorescence originating from two different Stark sublevels. Indeed, it is known that the energy level separations of the Stark sublevels is on the order of the thermal energy, and that the thermal occupation of the level above either the ground state or a metastable excited state levels can be expressed using the Boltzmann distribution. The fluorescence intensity ratio for these thermally coupled energy levels depends only on the energy level separation and on the temperature of the sample. This method guarantees a resolution of 1 °C, does not require high-resolution spectrometers, and can be used in real time [29, 31].

Both methods were tested and employed during the laser cooling experiments, and the temperatures obtained using them agree within experimental error. Details of these methods, as well as the complete experimental setups, can be found in [28, 31, 32].

## 3.4 Spectroscopic Analysis

BYF and YLF are anisotropic crystals and hence more than one polarized spectrum is needed to fully characterize them. In the case of $Yb^{3+}$ ions, in contast to $Tm^{3+}$ ions that are dominated by the electric-dipole term, the contribution of the magnetic term to transitions between different levels is not negligible, so six different spectra are required to fully characterize the Yb:BYF sample, and three for the Yb:YLF and Tm:BYF samples. In our case, we limited the study of the absorption spectra to the polarizations that coincide with the orientations used in the laser cooling experiments. In contrast, emission spectra were recorded for all possible polarizations at room temperature, since they are all required to estimate the mean average emission wavelength (see below).

The absorption spectra at room temperature did not show any spurious peaks due to unwanted impurities within the range of our instruments (0.3 – 3 µm), such as the divalent form of the ytterbium ion, which has strong broad absorption bands at around 400 and 650 nm. Figures 3.3 and 3.4 report the absorption spectra of the investigated crystalline samples. It is worth emphasizing the presence of several bands in these spectra. In fact, unlike in the case of ZBLAN glass, this transition can be exploited to pump the sample more efficiently in a laser cooling experiment, thus resulting in a larger overall efficiency.

Figure 3.3 presents the absorption band due to the transition $^2F_{7/2} \rightarrow\ ^2F_{5/2}$ of the $Yb^{3+}$ ion for the two investigated orientations of Yb:BYF and Yb:YLF. The strongest absorption peaks are $1.04 \times 10^{-20}$ cm$^2$ at 959 nm ($E \parallel c, H \parallel a$) and $0.79 \times 10^{-20}$ cm$^2$ at 959.5 nm ($E \parallel b, H \parallel a$) for the Yb:YLF and Yb:BYF crystals, respectively.

Figure 3.4 shows the absorption coefficients of 1.2% Tm:BaY$_2$F$_8$ for the $E \parallel b$ and $E \parallel c$ polarizations. These spectra show that the absorption range of Tm:BYF spans from 1600 nm up to ~ 2000 nm, with a main peak of $6.45 \times 10^{-21}$ cm$^2$ at 1642.2 nm for $E \parallel b$ polarization. From these spectra, it is clear that the absorption cross-section of $^3H_6 \rightarrow\ ^3F_4$ is smaller than that of the transition involved in laser cooling of Yb ions. This implies that a smaller fraction of the pump power will be absorbed by the crystal, and it also makes it more difficult to accurately determine the absorption spectra at wavelengths commonly used in laser cooling experiments, largely due to the fact that the absorption coefficient is very low. However, in order to predict the expected temperature drop in cooling measurements, the value of the absorption coefficient in the long-wavelength tail must be known to the highest precision. One way to overcome this problem is to estimate the absorption spectra from the emission spectra.

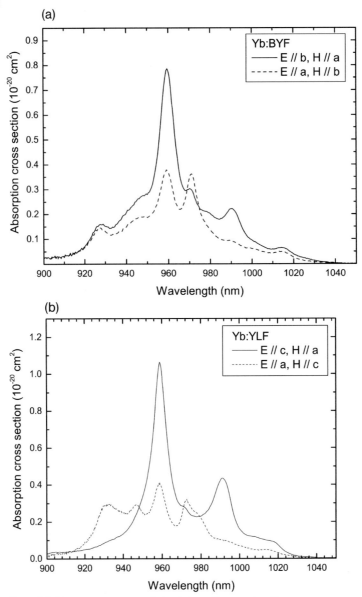

**Figure 3.3** Room-temperature polarized absorption spectra of 2.5% Yb:BYF (a) and 5% Yb:YLF (b).

Figure 3.5 shows the polarized emission spectra obtained with our experimental conditions. Even though the figures report only some of the possible polarizations for the sake of clarity, all of the possible polarizations were acquired and normalized relative to the absorbed power using a special mount, in order to ensure that the same experimental conditions were employed for all orientations of the sample.

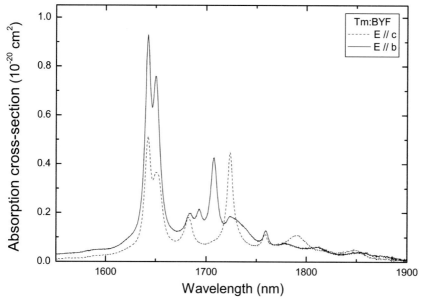

**Figure 3.4** Polarized absorption spectrum of 1.2% BaY$_2$F$_8$:Tm.

By using the generalized Einstein relations connecting the rate of spontaneous emission by the ions in the crystals to their rate of radiation absorption [27], it is possible to estimate the absorption spectra of the dopant ion from the emission spectra. Indeed, following [4], it is possible to write

$$\alpha_{abs}^{\sigma}(\lambda) \propto \lambda^5 I^{\sigma}(\lambda) e^{hc\lambda/kT}, \tag{3.1}$$

where $\alpha_{abs}^{\sigma}(\lambda)$ is the absorption coefficient at wavelength $\lambda$ for a polarization $\sigma$, $I^{\sigma}(\lambda)$ is the intensity of fluorescence emission for this polarization, and $T$ is the temperature of the sample. Using this equation, and starting from the emission spectra of 5% Tm:BYF, we derived the absorption cross-sections. The data obtained were then normalized to the absorption coefficient of the spectrometer at large absorption – over an order of magnitude above the noise level of our instrument. As shown in Figure 3.6, the data obtained with the reciprocity method agree with the spectrometer spectrum and give a better signal-to-noise ratio for $\alpha < 10^{-1}$ cm$^{-1}$.

Luminescence decay analysis showed that the profile is single exponential in form for all of the samples, excluding the existence of nonradiative energy transfer channels that may degrade the quantum efficiency and so the cooling efficiency. In the case of Yb-doped crystals, these measurements also clearly show the presence of multiple reabsorption and emission steps due to total internal reflection [28]. It is worth noting that, since these artifacts depend on the sample geometry and its refractive index, they cannot be avoided without coating the samples and thus increasing the complexity of the system. A clear way to overcome this problem is to choose a host that has a low index of refraction and contains a minimum number of absorbing impurity centers. The presence of these centers increases the

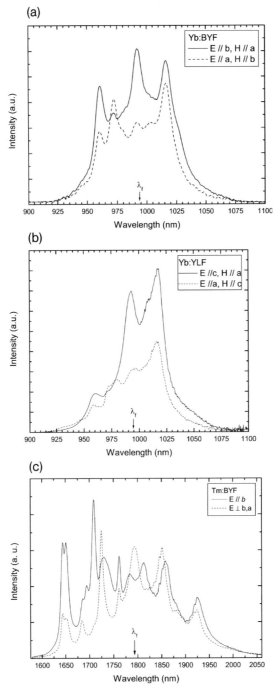

**Figure 3.5** Two of the polarized emission spectra of 2.5% Yb:BYF (a), 5%Yb:YLF (b) and 1.2% Tm:BYF (c).

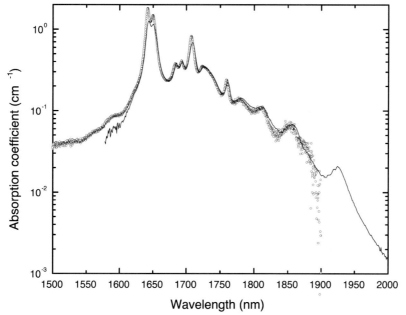

**Figure 3.6** $E \parallel b$-polarized absorption spectrum of 1.2% Tm:BaY$_2$F$_8$. The *open circles* represent the data obtained with the spectrometer, while the *solid line* indicates the absorption derived from the emission spectrum using the reciprocity method.

probability of nonradiative decay during the multiple emission–reabsorption steps, limiting the extrinsic quantum efficiency of the sample.

## 3.5 Cooling Results

### 3.5.1 Cooling Potential

The basic processes involved in laser cooling have been extensively studied by several authors [1,3,8], and we will only summarize the main concepts here. By pumping an Yb$^{3+}$- or Tm$^{3+}$-doped sample on the long-wavelength tail of the absorption spectrum, the ions are moved from the top of the ground state to the bottom of the excited state. The ions then thermalize within the manifold by rapid ($\tau \sim 10^{-12}$ s) absorption of phonons from the host and decay again to the ground state through spontaneous emission ($\tau \sim 10^{-3}$ s). In the ideal case, each fluorescent photon removes, on average, an amount of energy equal to the difference between the pump photon $h\nu$ and the mean fluorescent photon $h\nu_f$, defined as [30]

$$h\nu_f = \frac{\int h\nu \Phi^\sigma(\nu)\,d\nu}{\int \Phi^\sigma(\nu)\,d\nu}, \qquad (3.2)$$

where $\Phi^\sigma(\nu)$ is the emitted photon flux density (units of number per unit time per unit frequency interval) for a polarization $\sigma$.

By defining cooling efficiency as the ratio of the cooling power to the absorbed power, one obtains:

$$\eta_{\text{cool}} = \frac{P_{\text{cool}}}{P_{\text{abs}}} = \frac{h\nu_f - h\nu}{h\nu} = \frac{\lambda - \lambda_f}{\lambda_f} \approx \frac{kT}{h\nu}. \qquad (3.3)$$

Equation 3.3 states that if the material is pumped at longer wavelengths, the efficiency increases. It is clear, however, that moving further into the absorption tail will result in lower absorption, decreasing the absorbed power. For this reason, a more useful figure of merit for the cooling process should also take into account the absorption cross-section $\sigma_{\text{abs}}(\lambda)$ at the pumping wavelength. If we define the wavelength-dependent quantity $F_{\text{cool}}$ as the product of the cooling efficiency times the absorption cross-section:

$$F_{\text{cool}}(\lambda) = \sigma_{\text{abs}}(\lambda)\eta_{\text{cool}} = \sigma_{\text{abs}}(\lambda)\frac{\lambda - \lambda_f}{\lambda_f}, \qquad (3.4)$$

we see that this quantity is positive for $\lambda > \lambda_f$, may have one or more peaks (depending on the structure of the absorption spectrum), and tends to zero for $\lambda \gg \lambda_f$, when the absorption becomes negligible. The maximum of $F_{\text{cool}}$ can therefore be used to evaluate the cooling potential of a host. In a more realistic model, one should also take into account the percentage of radiation that is trapped in the crystal due to total internal reflection and the losses related to scattering, impurities and nonradiative decay (see for instance [8] for a more refined model). Nevertheless, this simple model provides useful hints when searching a host for laser cooling, since $F_{\text{cool}}$ can easily be calculated from absorption and emission spectra.

Table 3.2 summarizes the maximum cooling figures of merit for the hosts employed in laser cooling experiments.

**Table 3.2** Cooling figures of merit $F_{\text{cool}}$ of the hosts that have shown bulk cooling.

| Material | $F_{\text{cool}}$ ($10^{-23}$ cm$^2$) | Ref. |
|---|---|---|
| Yb$^{3+}$:YAG | 3.2 | [4] |
| Yb$^{3+}$:YSO | 1.5 | [4] |
| Yb$^{3+}$:BYF ($E \parallel b$, $H \parallel a$) | 1.4 | This work |
| Yb$^{3+}$:YLF ($E \parallel b$, $H \parallel a$) | 1.4 | This work |
| Tm$^{3+}$:BYF ($E \parallel b$) | 1.3 | This work |
| Tm$^{3+}$:ZBLAN | 1.1 | [9] |
| Yb$^{3+}$:ZBLAN | 0.7 | [3] |

This table shows that the maximum $F_{cool}$ values for Yb:BYF and Yb:YLF are $1.4 \times 10^{-23}$ cm$^2$ at 1015 and 1018 nm, respectively, and $1.3 \times 10^{-23}$ cm$^2$ at 1858 nm for Tm:BYF. These values are lower than those observed for oxide crystals, but are almost twice the value of $0.7 \times 10^{-23}$ cm$^2$ reported for Yb:ZBLANP glass, and are also greater than that of Tm:ZBLAN. The higher values of $F_{cool}$ can be ascribed to the higher absorption cross-sections of the crystals with respect to those of glasses, confirming the potential of this class of materials.

Finally, it is interesting to note that despite the fact that (3.3) seems to suggest that Tm$^{3+}$ ions (which have a smaller splitting between the ground and the first excited state) should be more efficient for laser cooling, Table 3.2 shows that their figure of merit is comparable with that of Yb$^{3+}$ ions. This can be ascribed to the fact that the absorption cross-section of Tm ions at the pumping wavelength is smaller than that of Yb ions. Therefore, from a theoretical point of view, it appears that the claimed advantages of Tm ions over Yb ions are compensated for by the fact that it is more difficult to effectively pump the sample. Moreover, since laser sources operating at wavelengths around 1 µm are currently more readily available, powerful and reliable than those operating around 2 µm, it soon becomes clear as to why Yb-doped materials are currently the most investigated host in laser cooling applications.

### 3.5.2
### Bulk Cooling

Figure 3.7 shows the typical temporal evolution of cooling after the sample is irradiated. The change in temperature is clearly exponential in form, as expected from simple thermodynamic considerations [33].

A maximum temperature drop of about 4 K starting from room temperature was achieved in single-pass configuration when the Yb:BYF was irradiated with 3 W at 1024 nm, the shortest available wavelength with our experimental setup, while a temperature drop of 6 K was observed in Yb:YLF when irradiated with 2.6 W at 1026 nm. The full potential of the Yb$^{3+}$-doped samples is still to be determined, since we could not scan the whole cooling wavelength range as we were limited by the restricted tunability range of the laser diodes. In fact, according to [8], the maximum temperature drop is located at a longer wavelength than that of the maximum $F_{cool}$ (1015 nm), with the exact value depending largely on the impurities in the sample. By pumping the crystal at a wavelength between 1015 and 1024 nm, we can expect to obtain a larger temperature drop.

In the case of Tm:BYF, the pump source was a homemade optical parametric oscillator [32]. This enabled us to scan the whole absorption region of interest for Tm ions – from 1800 to 2050 nm – as shown in Figure 3.8. This figure shows that a maximum temperature drop of ~ 7 W occurs when the sample is irradiated with 4.4 W at 1855 nm.

The importance of these results becomes clearer when the cooling efficiency obtained with fluoride single crystals is considered. The cooling power extracted may be estimated if we consider that it must balance the incoming power at equilibri-

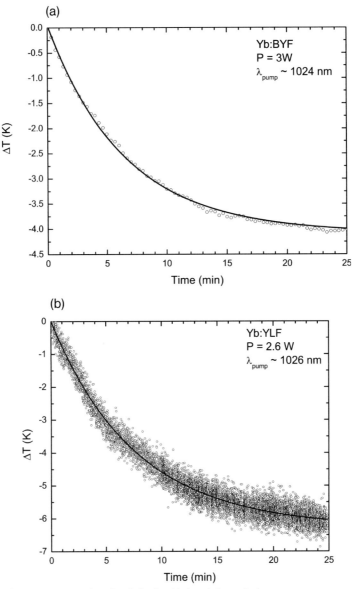

**Figure 3.7** Temporal cooling behavior (*dot*) and theoretical exponential fit (*line*) for 2.5% Yb:BYF (measured using a thermal camera; (a)) and 5% Yb:YLF (measured using a spectroscopic method; (b)).

um conditions, and that the heat loading on the sample is only due to the radiation from the vacuum chamber. We can then write:

$$P_{cool} = \varepsilon_s \sigma A \left( T_c^4 - T_s^4 \right), \tag{3.5}$$

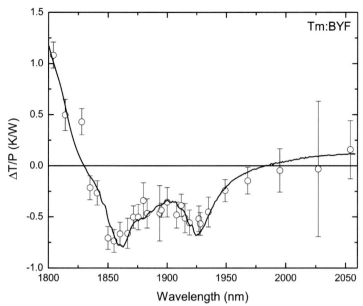

**Figure 3.8** Wavelength-dependent temperature change normalized to the pump power for BYF doped with 1.2% Tm for $E \parallel b$. Data points below the horizontal reference line indicate net cooling. The solid curve is a fit, as described in [32].

where $T_c$ and $T_s$ are the temperatures of the vacuum chamber and the sample, respectively, $A$ is the area, $\varepsilon_s$ is the emissivity of the sample, and $\sigma$ is the Stefan–Boltzmann constant. Using (3.5), cooling efficiencies of ~ 3% for the BYF crystals and ~ 2% for Yb:YLF have been estimated.

Figure 3.9 reports both the theoretical and the experimental cooling efficiencies versus the pumping wavelength for the 1.2% Tm:BYF crystal when pumped with the $E \parallel b$ polarization. A maximum cooling efficiency of $\eta_{max} = 3.4\%$ has been observed at 1934 nm. These results are comparable with, and even slightly better than, those of Tm:ZBLAN and YB:ZBLAN in the single-pass configuration, confirming that BYF is a valid alternative to the fluorozirconate glasses. The discrepancy between the theoretical and the experimental data at wavelengths longer than ~ 1950 nm can be ascribed to the very low absorption coefficient (cf. the large error bars). Note that the maximum cooling efficiency occurs at 1934 nm, while the maximum temperature drop occurs at 1855 nm. This follows from the definition of $\eta_{cool}$ as the ratio between the cooling and the absorbed power (see (3.3)): in the ideal case without impurities, the cooling efficiency increases with the wavelength. This is clearly visible in Figure 3.9 for wavelengths up to 1934 nm, beyond which the impurity absorption dominates and the cooling efficiency decreases. In contrast, the temperature drop is proportional to the cooling power, which is the product of the absorption coefficient and the cooling efficiency. However, this product is proportional to the figure of merit $F_{cool}$; indeed, we note that the maximum temperature

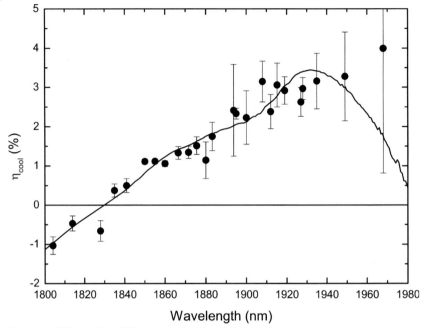

**Figure 3.9** $E \parallel b$ cooling efficiency as a function of the pump wavelength for the 1.2% $Tm^{3+}$-$BaY_2F_8$ crystal. Theoretical prediction (*solid line*) and experimental data (*dots*).

**Table 3.3** Temperature drops and cooling efficiencies observed in crystalline hosts and in ZBLAN (for comparison). All of the values were measured in the single-pass configuration and in bulk samples, except for the 16 K temperature drop of Yb:ZBLAN, which was observed in a fiber sample (n.d. = value not declared).

| Material | $\Delta T$ (K) | $\eta_{cool}$ (%) | Ref. |
| --- | --- | --- | --- |
| $Yb^{3+}$:YAG | 0.36 | 2 | [4] |
| $Yb^{3+}$:YSO | 1 | n.d. | [4] |
| $Yb^{3+}$:BYF | 4 | 3 | This work |
| $Yb^{3+}$:YLF | 6 | 2 | This work |
| $Tm^{3+}$:BYF | 6 | 3 | This work |
| $Tm^{3+}$:ZBLAN | 1.2 | n.d. | [8] |
| $Yb^{3+}$:ZBLAN | 16 | 2 | [34, 35] |
| $Er^{3+}$:$KPb_2Cl_5$ | 0.5 | 0.682 | [11] |

drop occurs at a wavelength close to the one where $F_{cool}$ is maximal (1858 nm; see Section 3.5.1).

Table 3.3 summarizes the results obtained in the single-pass configuration and compares them with those from different hosts. More interesting than the maximum temperature drop obtained – which depends on the pump power and the geometry of the sample (the 16 K temperature drop in Yb:ZBLAN was observed in a fiber-shaped sample, for instance) – is the value of the cooling efficiency. This table indicates that the cooling efficiency of BYF crystals is larger than the ~ 2% efficiency reported for both Yb:YAG [4] and Yb:ZBLANP [35], confirming that this material could play an important role as a competitor to the fluorozirconate glasses.

In the case of Yb:YLF, the slightly lower ~ 2% efficiency may be ascribed to multiple absorption–re-emission processes. It is well known that part of the light emitted in a fluorescent solid is trapped inside the body due to total internal reflection, and this phenomenon can easily be seen in fluorescence decay time measurements [28]. The trapped photons will be eventually reabsorbed by the dopant ions and further emission will occur. Anyway, each time a fluorescent photon is absorbed, it has a probability that is equal to the quantum efficiency of being re-emitted. Increasing the number of steps a photon needs to traverse to escape from the crystal will eventually result in a lower external overall quantum efficiency. Moreover, a photon traveling inside the crystal can transfer its energy directly to the host if it is absorbed by some impurities or if inelastic scattering processes occur. Clearly, the amount of light trapped depends on the dopant concentration and the refractive index of the crystal (a simple model that describes this effect and estimates the cooling efficiency can be found in [37]). This may therefore explain the smaller cooling efficiency observed in the 5% Yb:YLF crystal.

However, a higher dopant concentration allows the sample to be pumped more efficiently, and this fact may be exploited in a more refined pumping scheme in order to obtain a larger temperature drop. Indeed, a similar experiment was recently carried out using a cavity-enhanced resonant absorption technique [36]. By placing the sample inside a resonant cavity, it was shown that is possible to increase the pump power absorbed by the cooling sample up to about 90%. In this way, by directing a 15 W laser operating at 1030 nm onto 5% Yb:YLF, a temperature drop of ~ 70 K with respect to the surrounding environment has been attained, corresponding to an extracted power of 58 mW [38].

## 3.6 Conclusion

A spectroscopic investigation and an analysis of its optical and mechanical properties have confirmed the good properties of fluoride single crystals for laser cooling applications.

We observed bulk cooling in the single-pass configuration in all of the studied samples. The maximum temperature drops were almost 4, 6 and 7 K for 2.5% Yb:BYF, 5% Yb:YLF and 1.2% Tm:BYF, respectively, corresponding to a cooling efficiency of about 3% for all of the samples except 5% Yb:YLF, which has a cooling efficiency of about 2%. The lower efficiency of Yb:YLF, which is still comparable with that of Yb:ZBLAN, could be ascribed to multiple absorption–re-emission processes which increase with the $Yb^{3+}$ concentration and may reduce the quantum efficiency. However, the high absorption coefficient of this sample can be exploited to obtain large temperature drops.

## Acknowledgments

The authors wish to thank I. Grassini for preparing the samples, and H. P. Jenssen and A. Cassanho, who – due to their expertise in crystal growth, developed at AC Materials (Tampa, FL, USA) – provided precious suggestions. We thank A. Di Lieto, D. Parisi, L. Bonelli, F. Cornacchia and A. Toncelli for useful discussions.

## References

1. EPSTEIN, R.I., BUCHWALD, M.I., EDWARDS, B.C., GOSNELL, T.R. AND MUNGAN, C.E. (1995) Nature, 377, 500.
2. THIEDE, J., DISTEL, J., GREENFIELD, S.R. AND EPSTEIN, R.I. (2005) Appl. Phys. Lett., 86, 154107.
3. BOWMAN, S.R. AND MUNGAN, C.E. (2000) Appl. Phys. B, 71, 807.
4. EPSTEIN, R.I., BROWN, J.J., EDWARDS, B.C. AND GIBBS, A. (2001) J. Appl. Phys., 90, 4815.
5. EDWARDS, B.C., ANDERSON, J.E., EPSTEIN, R.I., MILLS, G.L. AND MORD, A.J. (1999) J. Appl. Phys., 86, 6489.
6. YATSIV, S. (1961) In Advances in Quantum Electronics, (ed J.R. Singer), New York, Columbia Univ. Press.
7. EDWARDS, B.C., BUCHWALD, M.I., EPSTEIN, R.I., GOSNELL, T.R. AND MUNGAN, C.E. (1996) In Proceedings of the 9th Annual American Institute of Astronautics & Aeronautics Utah State Conference on Small Satellites, F. Redd, Ed., Reston: American Institute of Astronautics & Aeronautics.
8. HOYT, C.W., SHEIK-BAHAE, M., EPSTEIN, R.I., EDWARDS, B.C. AND ANDERSON, J.E. (2000) Phys. Rev. Lett., 85, 3600.
9. HOYT, C.W., SHEIK-BAHAE, M., EPSTEIN, R.I., GREENFIELD, S., THIEDE. J., DISTEL, J. AND VALENCIA, J. (2003) J. Opt. Soc. Am. B, 20, 1066.
10. MENDIOROZ, A., FERNÁNDEZ, J., VODA, M., AL-SALEH, M. AND BALDA, R. (2002) Opt. Lett., 27, 1525.
11. FERNÁNDEZ, F., GARCIA-ADEVA, A.J. AND BALDA, R. (2006) Phys. Rev. Lett., 97, 033001.
12. JOHNSON, L.F. AND GUGGENHEIM, H.J. (1973) Appl. Phys. Lett., 23, 96.
13. JENSSEN, H.P. AND CASSANHO, A. (2006) In Proc. SPIE – Photonics West, 6100-24.
14. VANNINI, M., TOCI, G., ALDERIGHI, D., PARISI, D., CORNACCHIA, F. AND TONELLI, M. (2007) Opt. Express 15, 7994.
15. CORNACCHIA, F., RICHTER, A., HEUMANN, E., HUBER, G., PARISI, D. AND TONELLI, M. (2007) Opt. Express 15, 992.
16. TABIRIAN, A.M., JENSSEN, H.P. AND CASSANHO, A. (2001) Efficient, room temperature mid-infrared laser at 3.9 µm in Ho:BaY$_2$F$_8$, in: OSA Proc. Advanced Solid-State Lasers, Trends in Optics and Photonics 50, 170.
17. GALZERANO, G., CORNACCHIA, F., PARISI, D., TONCELLI, A., TONELLI, M. AND LAPORTA, P. (2005) Opt. Lett. 30, 854.
18. IZOTOVA, O.E. AND ALEKSANDROV, V.B. (1970) Sov. Phys. Dokl., 16, 525.
19. BENSALAH, A., GUYOT, Y., ITO, M., BRENIER, A., SATO, H., FUKUDA, T. AND BOULON, G. (2004) Opt. Mater., 26, 375.
20. TONCELLI, A., TONELLI, M., CASSANHO, A. AND JENSSEN, H.P. (1999) J. Lum., 82, 291.

21 ORLOVSKII, Y.V., BASIEV, T.T., VOROB'EV, I.N., ORLOVSKAYA, E.O., BARNES, N.P. AND MIROV, S.B. (2002) *Opt. Mater.*, **18**, 355.

22 WEBER, M.J. (2002) *Handbook of Optical Materials*, Boca Raton, CRC.

23 KAMINSKII, A.A. (1993) *Phys. Stat. Sol. A*, **137**, K61.

24 SHIKIDA, A., YANAGITA, H. AND TORATANI, H. (1994) *J. Opt. Soc. Am. B*, **11**, 928.

25 VODA, M., AL-SALEH, M., LOBERA, G., BALDA, R. AND FERNÁNDEZ, J. (2004) *Opt. Mater.* **26**, 359.

26 ALDERIGHI, D., TOCI, G., VANNINI, M., PARISI, D., BIGOTTA, S. AND TONELLI, M. (2006) *Appl. Phys. B* **83**, 51.

27 MCCUMBER, D.E. (1964) *Phys. Rev.* 4A, A954.

28 BIGOTTA, S., PARISI, D., BONELLI, L., DI LIETO, A., TONCELLI, A. AND TONELLI, M. (2006) *J. Appl. Phys.*, **100**, 013109.

29 WADE, S.A., COLLINS, S.F. AND BAXTER, G.W. (2003) *J. Appl. Phys.* **94**, 4743.

30 MUNGAN, C.E. AND GOSNELL, T.R. (1998) *Laser Cooling of Solids (Vol. 40 of Advances in Atomic, Molecular, and Optical Physics)*, New York, Academic.

31 BIGOTTA, S., DI LIETO, A., PARISI, D., TONCELLI, A. AND TONELLI, M. (2007) In *Proc. SPIE – Photonics West* 64610E.

32 Patterson, W., Bigotta, S., Sheik-Bahae, M., Parisi, D., Tonelli, M. and Epstein, R.I. (2008) *Opt. Express*, **16**, 1704.

33 CLARK, J.L., MILLER, P.F. AND RUMBLES, G. (1998) *J. Phys. Chem. A*, **102**, 4428.

34 MUNGAN, C.E., BUCHWALD, M.I., EDWARDS, B.C., EPSTEIN, R.I. AND GOSNELL, T.R. (1997) *Phys. Rev. Lett.*, **77**, 1030.

35 FAJARDO, J.C., SIEGEL, G.H. JR., EDWARDS, B.C., EPSTEIN, R.I., GOSNELL, T.R. AND MUNGAN, C.E. (1997) *J. Non-Cryst. Solids*, **213–214**, 95.

36 SELETSKIY, D., HASSELBECK, M.P., SHEIK-BAHAE, M., THIEDE, J. AND EPSTEIN, R.I. (2006) In *Advanced Optical and Quantum Memories and Computing III*, **6130**, 61300P, Bellingham, SPIE.

37 BIGOTTA, S. (2006) PhD Thesis. Pisa, Universitá di Pisa.

38 SELETSKIY, D., HASSELBECK, M.P., SHEIK-BAHAE, M., EPSTEIN, R., BIGOTTA, S. AND TONELLI, M. (2007) In *CLEO/QELS Conf.*, Baltimore, MD, USA, 8–10 May.

# 4
# Er$^{3+}$-Doped Materials for Solid-State Cooling
*Joaquin Fernandez, Angel Garcia-Adeva and Rolindes Balda*

## 4.1
## Low Phonon Energy Materials

The investigation of new hosts for rare-earth ions with low phonon energies appears to be a promising way to find efficient cooling materials, especially for dopant ions with low-energy band gaps between active levels. Hosts with low phonon energies lead to low nonradiative transition rates due to multiphonon relaxation and high radiative transition rates, which increases the quantum efficiency from excited states of active ions. Sulfide [1] and chloride [2] based hosts have been studied, as their phonon energies are lower than in fluoride matrices. Among chloride hosts, $KPb_2Cl_5$ crystal has become an interesting material due to its low maximum phonon frequency (203 cm$^{-1}$) [2], and the fact that it is not hygroscopic. Indeed, until a few years ago, no moisture-resistant chloride laser host had been identified. It incorporates RE ions, has a high chemical stability, and melts at a low temperature. However, $KPb_2Cl_5$ crystal presents poor mechanical properties and it is difficult to synthesize. $KPb_2Cl_5$ has been the subject of various crystal growth methods and studies concerning the luminescence mechanism of divalent lead emission centers [3, 4], its crystal structure, and its absorption and emission properties [5].

On the other hand, considerable progress has been made in the last two decades in both the discovery of new compositions for halide glasses and our knowledge of their optical properties. Interest in these materials is mainly due to the possibility of extending the infrared (IR) transparency domain towards longer wavelengths and consequently achieving mid-infrared ultratransparency. However, besides the potential uses related to their passive optical behavior, many other important applications arise when glasses are doped with rare-earth (RE) ions.

The optical properties of rare-earth ions in glasses depend on the chemical composition of the glass matrix, which determines the structure and nature of the bonds [6]. The development of new glass-based optical devices requires a better understanding of the interionic interactions that are deeply involved in the fundamental physics of rare-earth ions. Among the numerous halide glass compositions that have been described in the literature, only a few ones have been investigated in depth, because the study of their optical properties requires samples of a reasonable

*Optical Refrigeration. Science and Applications of Laser Cooling of Solids.*
Edited by Richard Epstein and Mansoor Sheik-Bahae
Copyright © 2009 WILEY-VCH Verlag GmbH & Co. KGaA, Weinheim
ISBN: 978-3-527-40876-4

size, optical quality, and good resistance to moisture corrosion. Most heavy metal fluoride based glasses fulfill these requirements and so have received increasing attention. In particular, because of their extended infrared transmission, a great effort to develop ultralow-loss optical fibers in the mid-infrared has been realized. On the other hand, their ability to incorporate significant amounts of RE ions and their high emission efficiency due to multiphonon emission rates, which are lower than in other glasses, have made them attractive candidates for optical amplifiers and laser applications [7]. Besides heavy metal fluoride glasses, there are a number of heavy metal halide systems that are known to transmit further in the infrared than pure fluoride glasses. $ZnCl_2$ is the best glass-former and the most widely known. Its infrared edge in the 12 to 13 µm region is of interest for optical operation in the 8 to 12 µm atmospheric window [8]. Although many attempts have been made to increase its resistance to water corrosion and to devitrification, its hygroscopicity is still very high and affects the optical properties in the IR region. Cadmium chloride has also been proven to lead to vitreous materials [9]. When $CdCl_2$ glasses are stabilized by a mixture of chloride and fluoride, such as in $CdF_2$-$BaF_2$-NaCl or others [10], the resistance to aqueous corrosion increases but the multiphonon edge is then determined by the metal fluoride bond and it shifts to shorter wavelengths. Recently, fluorochloride glasses based on Cd, Na, Ba and Zn have been obtained which are stable and present lower phonon energies (370 $cm^{-1}$) than those of conventional fluorozirconates or barium-indium-gallium fluoride glasses. The limit of infrared transparency in these fluorochloride glasses is around 9 µm, with a shift of 4 µm compared to well-known ZBLAN fluoride glasses [11]. Therefore, high emission efficiencies can be expected for rare earths in these matrices.

### 4.1.1
### $KPb_2Cl_5$ Crystal

$KPb_2Cl_5$ crystal is biaxial, crystallizes in the monoclinic system (space group $P2_1/c$) with lattice parameters $a = 0.8831$ nm, $b = 0.7886$ nm, $c = 1.243$ nm, and $\beta = 90.14°$, and is transparent in the 0.315 µm to 20 µm spectral region [4]. The $Pb^{2+}$ ions occupy two crystallographically nonequivalent positions differing in the number of neighboring chlorine ions and their positions. The first one, $Pb^{2+}(1)$, has an octahedral environment and a Pb–Cl distance of < 3 Å. The second one, $Pb^{2+}(2)$, assumes the same coordination that $Pb^{2+}$ has in $PbCl_2$ [5]. The refractive index was measured by means of ellipsometric measurements, giving a value of $n = 2.016 \pm 0.005$ [12].

Fluorescence line narrowing experiments have demonstrated the existence of three different local environments around the RE ions in $KPb_2Cl_5$. The RE ions may occupy both the Pb and the K sites, but the luminescence results suggest that RE ions are most likely to occupy the Pb(2) site [13].

Chlorides generally hydrolyze, and therefore crystals contain some products of hydrolysis. The most frequent radicals in KCl are anions like $(OH)^-$, $(O_2)^-$, and $(CO_3)^{2-}$. When heated, $PbCl_2$ not only hydrolyzes but it also decomposes. Moreover, if the material contains water molecules or other impurities, the decomposition increases. Due to the oxygen and moisture sensitivity of halide components of $KPb_2Cl_5$ and also RE-doped halides, this requires very the strict exclusion of both $O_2$ and $H_2O$ from the system. For this reason, single crystals of chloride compounds for laser cooling experiments have been grown mainly by the Bridgman–Stockbarger method, using sealed silica ampoules.

In the $KCl$-$PbCl_2$ system, a congruent compound, $KPb_2Cl_5$, is formed at the 1 : 2 stoichiometric molar ratio of the components [14, 15].

On cooling down, a phase transition was found at 251 °C [15] or 270 °C [5]. As mentioned above, solid halides such as KCl, $PbCl_2$, and $RECl_3$ (RE = $Yb^{3+}$, $Pr^{3+}$, $Nd^{3+}$, $Er^{3+}$, $Eu^{3+}$) readily absorb $H_2O$ when exposed to moisture. At elevated temperatures and in the molten state, the presence of $O_2$ and $H_2O$ facilitates the formation of lead hydroxychloride and various oxide chlorides, which produce cloudiness in the melt, and the resulting crystals are colored and present cracks because of the lattice distortion and scattering centers, which makes them useless.

The major problems encountered in the growth of halide compounds (single and ternary) with a high optical quality are thus: the need for very strict exclusion of oxygen impurities from the starting materials and for the prevention of any contact between the purified material and the air (i.e. oxygen and water) during both handling operations for the starting materials and the single-crystal growth process.

The starting materials for the growth of pure and doped $KPb_2Cl_5$ crystals were high-purity $PbCl_2$ puratronic 99.999%, KCl puratronic 99.99%, and $ErCl_3$ 99.99%, all supplied by Alfa Aesar, Johnson Matthey Company.

By drying $PbCl_2$ and also KCl in vacuum, it is only possible to reduce the humidity concentration [16, 17]. Therefore, before the crystal growth is initiated, moisture and all oxide impurities must be removed using other procedures. One effective purification method for lead and potassium chlorides is the chlorination of the melts with a mixture of highly purified argon (or another inert gas) and $CCl_4$ [16–21] in combination with subsequent horizontal zone-refining. The Ar-$CCl_4$ atmosphere has chlorinating properties and also diminishes the excess of lead and oxide content. We applied this method to purify both KCl and $PbCl_2$ using a quartz vessel. The products were dried beforehand under vacuum ($10^{-2}$ mm Hg) at 220 °C until a sharp drop in pressure was observed. To remove the remaining content of oxygen, water, and $OH^-$ molecules, a stream of built-in purified argon gas (< 20 ppb of $H_2O$ and < 10 ppb of $O_2$) containing $CCl_4$ (99.9%) was flushed into the content of the ampoules, and the temperature was increased from room temperature up to 50 °C above the melting temperature of each compound (550 °C for $PbCl_2$, 820 °C for KCl, and 484 °C for $KPb_2Cl_5$). The stream of argon-$CCl_4$ was introduced at the bottom of the ampoule containing the substance by means of a fused silica capillary tube, as shown in Figure 4.1.

By further heating of the furnace, the oxygen and eventually water vapor traces are eliminated by the pyrolyzed $CCl_4$ [18] in the form of volatile compounds like HCl, CO or $CO_2$. A successful chemical reaction was achieved when the meniscus of the melt changed and became convex, in contrast to the concave menisci (strong wetting of the wall) of nonchlorinated melts. After the reaction is conclud-

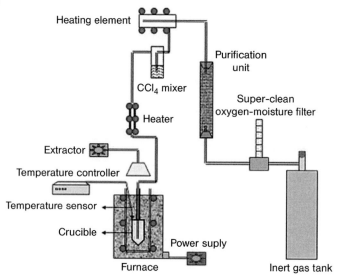

**Figure 4.1** Experimental arrangement for bubbling the raw materials in order to grow ternary lead chloride compounds.

ed, the capillary tube is withdrawn and the ampoule is sealed off by making use of an oxygen-methane blowtorch. After this treatment, the ampoule is slowly cooled down to room temperature.

The purification process is continued by zone-refining in a one-ring resistance horizontal furnace in order to heat a narrow zone. The total number of zone passes went from five to ten. The resulting compounds were transparent and colorless and did not adhere to the walls of the quartz ampoules.

The growth was performed using $PbCl_2$ and KCl prepared by the method mentioned before. A molar ratio of 66.7 mol % of $PbCl_2$ and 33.3 mol % of KCl was mixed and loaded into a quartz ampoule. A new drying period under vacuum was followed by bubbling with $CCl_4$ carried by argon. The quartz ampoule containing the melt was closed under a chlorine atmosphere and moved into the furnace for crystal growth by the Bridgman vertical lowering technique.

The crystal growth was performed in a two-zone transparent vertical furnace with a Kanthal wire as the heating element, which could operate with different temperature gradients by using two independent temperature controllers.

The regulation and accurate control of the temperature profile and gradient are of great importance to the result and the quality of the growing process. We ensure that temperature is controlled by utilizing a thyristor unit from Eurocube, model 425 (220 V, 40 A), which was controlled in a single phase load mode by a signal originating from a Eurotherm 818P controller. The temperature measured at the point where crystallization occurs was maintained to within 0.1 °C.

Transparent furnaces have also been used by some other authors for the growth of single crystals, like transition metal doped $CdCl_2$ [21], $PbCl_2$ [19], and $CaGa_2S_4$ doped with $Dy^{3+}$ [22]. This type of furnace has the advantage of facilitating observations of the solid–liquid interface at various translation velocities and temperature gradients. We performed crystal growth using different temperature gradients and a lowering speed of $1-2\,\text{mm}\,\text{h}^{-1}$. After completing the growth, the crystals were cooled down to room temperature at a rate of $2-4\,°\text{C}\,\text{h}^{-1}$. The resulting single crystals were water-white, optically transparent, and free of twins and cracks. Their diameters ranged between 12 and 16 mm, and lengths ranged from 20 to 40 mm.

The Er-doped $KPb_2Cl_5$ crystals were grown by the same procedure. The dopant concentrations were situated in the $x = 0.05–2.0$ mol % range in the melt. The three purified components ($KCl$, $PbCl_2$, and $ErCl_3$) were loaded together into the quartz cell at a molar ratio of $1:2:x$ for bubbling with Ar-$CCl_4$ mixture, and the process was repeated, just as for the undoped crystals. The quality of the resulting crystals was maintained and they were uniformly colored.

### 4.1.2
### Fluorochloride Glasses

Divalent chlorofluoride glasses of composition $22CdCl_2$-$18CdF_2$-$30NaF$-$20BaF_2$-$10ZnF_2$ doped with $Er^{3+}$ were prepared at the Laboratoire de Verres et Céramiques of the University of Rennes (France). Samples were synthesized using conventional methods with reagent-grade fluorides and chlorides under a dry argon atmosphere. The glass transition temperature, $T_g$, and crystallization temperature, $T_x$, were measured by differential scanning calorimetry (DSC). The glass-forming ability given by the $\Delta T$ parameter, defined as $\Delta T = T_x - T_g$, is $101 \pm 4\,°\text{C}$. This non-hygroscopic divalent fluorochloride glass has lower phonon energies ($370\,\text{cm}^{-1}$) than ZBLAN fluoride glasses due to the lower fundamental energies of Cd–Cl bonds ($250\,\text{cm}^{-1}$) and Cd–F bonds ($370\,\text{cm}^{-1}$).

## 4.2
## Internal Cooling Measurements

In order to probe local or internal cooling, collinear photothermal deflection spectroscopy is commonly used. This technique is based on the mirage effect, where a probe laser beam probes the thermal lens created by a pump laser beam. When the pump laser is adequately tuned so that we enter the cooling region, there is a change in the direction in which the probe beam propagates. This shows up in the corresponding spectrum as a 180° change of the signal phase. Also, the magnitude of the signal goes to zero upon moving from the heating to cooling region (and vice versa); that is, when the barycenter of the energy band is crossed. A schematic diagram of the setup used to perform these experiments can be seen in Figure 4.2, together with a picture of the actual setup employed to collect the data reported below. The beam of a tunable ($\lambda = 780-910$ nm) cw Ti:sapphire ring laser (8 GHz bandwidth) that enters the sample perpendicularly to the center of its faces is modulated at 1.24 Hz by means of a mechanical chopper. A fraction of the incident power is used for signal normalization. A copropagating He-Ne probe laser beam

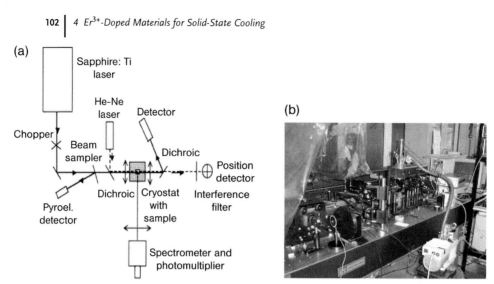

**Figure 4.2** On the *left*, a block diagram of the setup used for the collinear photothermal deflection spectroscopy measurements; on the *right*, a picture of the actual setup for this technique.

($\lambda$ = 632.8 nm) is coaligned with the pump beam through a dichroic element. Both pump and probe copropagating beams are focused into the middle of the sample with diameters of ~ 100 and ~ 60 µm, respectively. After leaving the sample, the beams pass through a second identical lens separated from the first one by a distance of twice the focal length (5 cm) in order to avoid high divergence of the emerging beams. A second dichroic beam splitter deviates the pumping beam into a pyroelectric detector that measures the transmitted pumping power. The probe beam passes through an interference filter to eliminate residual pumping radiation before reaching a quadrant position detector. This same configuration also allows emission and excitation spectra to be measured by simply collecting the fluorescence perpendicular to the focused area of the pumping beam using a collimating lens, and focusing it with a second lens at the 100-µm entrance slit of a 0.22-m monochromator equipped with an extended infrared photomultiplier. Lock-in detection is used in both experiments. Thermal deflection waveforms are detected using a digital scope.

One nice feature of this technique is that it allows the cooling/heating efficiencies to be easily estimated from the slopes of the PDS signal magnitude around the barycenter of the emitting band. The evaluation of the QE is carried out by considering a simplified two-level system for each of the transitions involved. In this model, a laser photon of angular frequency $\omega_L$ from the incident beam of intensity $I_0$ modulated at an angular frequency $\omega_m$ is absorbed by an electron that is promoted to the excited state. The relaxation to the ground state can take place through radiative or nonradiative processes with probabilities of $W_{rad}$ and $W_{nr}$, respectively, at a mean energy $\hbar\omega_0$. The energy difference between the incident and fluorescent photons is exchanged as heat with the host. In this model, the heat the sample exchanges per unit time and unit volume in a typical heating process is given by

$$H = N_1[W_{nr}\hbar\omega_L + W_{rad}\hbar(\omega_L - \omega_0)], \quad (4.1)$$

where $N_1$ is the population density of the excited state. The excited state population is governed by the incident beam modulation frequency $\omega_m$ and the lifetime of the excited state, $\tau$,

$$\frac{dN_1}{dt} = N_0 \frac{\sigma I_0}{\hbar\omega_L} - \frac{N_1}{\tau}, \quad (4.2)$$

where $\sigma$ is the absorption cross-section and $N_0$ is the population density of the ground state. If we solve this equation (in the steady state) and enter the result into (4.1), we find the following value for the heat exchanged with the sample:

$$H(\omega_L, \omega_m) = \frac{I_0 N_0 \sigma}{2} \cos\Phi \sin(\omega_m t + \Phi)\left(1 - \eta\frac{\omega_0}{\omega_L}\right), \quad (4.3)$$

where $\Phi = \tan^{-1}(\omega_m \tau)$ is the phase relative to the incident modulation beam. For a small modulation frequency, $\tau\omega_m \ll 1$, the heat exchanged with the sample per unit time and unit volume takes the very simple form

$$H(\omega_L) = N_0 \sigma I_0 \left[1 - \eta\frac{\omega_0}{\omega_L}\right], \quad (4.4)$$

where the fluorescence quantum efficiency of the excited state is defined as the ratio

$$\eta = \frac{W_{rad}}{W_{rad} + W_{nr}} = W_{rad}\tau. \quad (4.5)$$

In the collinear configuration, the amplitude of the angular deviation of the probe beam is always proportional to the amount of heat the sample exchanges, whatever its optical or thermal properties are. This allows the QE of the transition to be obtained from the ratio of the photothermal deflection amplitude (PDS) to the sample absorption (Abs) obtained as a function of the excitation wavelength $\lambda$ around the mean fluorescence wavelength $\lambda_0$

$$\frac{PDS}{Abs} = C\left(1 - c\frac{\lambda}{\lambda_0}\right), \quad (4.6)$$

where $C$ is a proportionality constant that depends on the experimental conditions. The mean fluorescence wavelength above which cooling is expected to occur needs to be calculated by taking into account the branching ratios for the emissions from the emitting manifold. The calculated value is close to the one found experimentally for the transition wavelength at which the cooling region begins [23].

A typical normalized photothermal deflection spectrum around the zero deflection signal (852.5 nm) can be seen in Figure 4.3a for the $Er^{3+}$:$KPb_2Cl_5$ sample. It

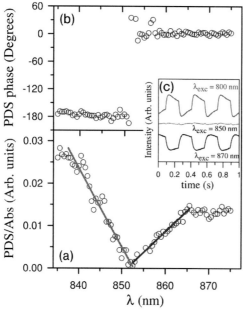

**Figure 4.3** (a) Signal deflection amplitude normalized by the sample absorption as a function of pumping wavelength for the $Er^{3+}$:$KPb_2Cl_5$ crystal. (b) Phase of the photothermal deflection signal as a function of pumping wavelength. (c) Photothermal deflection signal waveforms in the heating (800 nm) and cooling (870 nm) regions and around the cooling threshold (850 nm).

was measured at an input power of 1.5 W. The least square fits to expression (4.6) in both the cooling and heating regions are displayed. The resulting QE values are $0.99973 \pm 0.00008$ and $1.00345 \pm 0.00004$, respectively, and therefore the cooling efficiency estimated via the QE measurements is $0.37 \pm 0.01\%$. As predicted by the theory [24], a sharp jump of 180° in the PDS phase measured by lock-in detection can be observed during the transition from the heating to the cooling region (see Figure 4.3b). Figure 4.3c shows the PDS amplitude waveforms registered in the oscilloscope at three different excitation wavelengths: 800 (heating region), 852.5 (mean fluorescence wavelength), and 870 nm (cooling region). It is apparent that at 852.5 nm the signal is almost zero, whereas in the cooling region, at 870 nm, the waveform of the PDS signal shows an unmistakable phase reversal of 180° when compared with the one at 800 nm.

Figure 4.4 shows the results for the $Er^{3+}$:CNBZn glass (obtained at a pump power of 1.9 W), where the zero deflection signal occurs around 843 nm. The 180° change of the PDS phase is also clearly attained but with a little less sharpness than for the $Er^{3+}$:$KPb_2Cl_5$ crystal (see Figure 4.4b). The QE values corresponding to the heating and cooling regions are $0.99764 \pm 0.00005$ and $1.00446 \pm 0.00001$, respectively, and the estimated cooling efficiency is $0.682 \pm 0.006\%$. The PDS waveforms corresponding to the heating and cooling regions are shown in Figure 4.4c. It is worth

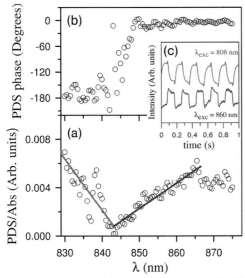

**Figure 4.4** (a) Signal deflection amplitude normalized by the sample absorption as a function of pumping wavelength for the $Er^{3+}$:CNBZn glass. (b) Phase of the photothermal deflection signal as a function of pumping wavelength. (c) Photothermal deflection signal waveforms in the heating (808 nm) and cooling (860 nm) regions.

noting that the cooling processes in both systems can be obtained at quite low power excitations. As an example, for the $Er^{3+}$:$KPb_2Cl_5$ crystal, cooling by anti-Stokes emission (CASE) is still efficient at a pump power of only 500 mW.

## 4.3
## Bulk Cooling Measurements

The results described in the previous section demonstrate that these Er-doped low phonon energy materials are capable of internal laser cooling in a certain spectral range, even at small pumping powers. However, more important from a practical point of view is the question of whether these materials would yield bulk or macroscopic cooling. To explore this issue, one usually employs a thermal infrared camera that is able to map the temperature field of the sample under study.

We conducted such a study by measuring the absolute temperature of these materials as a function of time for several pumping powers between 0.25 and 1.9 W. Also, a range of wavelengths that sweeps across the heating and cooling regions described above was explored in order to quantitatively assess the cooling potential of these Er-doped materials. To perform these measurements, a Thermacam SC 2000 (FLIR Systems) infrared thermal camera was used. This camera operates between object temperatures of −40 and 500 °C with a precision of ±0.1 °C. The detector is an array of 320 × 240 microbolometers. The camera is connected to an acquisition card interface that is able to record thermal scans at a rate of 50 Hz. The absolute temperature was calibrated with a thermocouple located at the sample holder. A picture of the experimental setup can be seen in Figure 4.5. Thermal scans at a rate of one image per second were acquired for time intervals that depend on the particular data series. The camera was placed 12 cm apart from the window cryostat so that a lens with a vision field of 45° allows the camera to be focused on the sample. Figure 4.6 shows the runs performed at 870 and 860 nm for the $Er^{3+}$:$KPb_2Cl_5$ crystal and $Er^{3+}$:CNBZn glass samples, respectively, using the same pump geometry conditions as the ones described above. According to the PDS measurements reported above, these pumping wavelengths are well inside the cooling region for both materials. The laser power on the sample was fixed at 1.9 W in both cases. The insets in that figure depict some examples of the thermal scans obtained with the infrared camera. It is clear from those colormaps that the sample is cooling down over time. However, it is difficult to extract any quantitative information about the cooling of the sample, as these changes are small compared with the absolute temperature of its surroundings. For this reason, in order to assess whether cooling is occurring in the bulk, we calculated the average temperature of the area enclosed in the green rectangles depicted in the upper insets in Figure 4.6, and the corresponding results constitute the green curves in that figure. Both samples clearly cool down under laser irradiation. The $Er^{3+}$:$KPb_2Cl_5$ sample temperature drops by 0.7 ± 0.1 °C in 1500 s. To check that this temperature change was indeed due to laser cooling, the laser was turned off at that point. This can be easily identified as an upturn in the curve that represents the evolution of the sample temperature, which means that this quantity starts to rise as soon as the laser irradiation is stopped. On the other hand, the temperature of the $Er^{3+}$:CNBZn glass sample starts to rise when laser irradiation starts. After ~ 150 s, this tendency is inverted and the sample starts to cool down. From that point on (and in approximately 1000 s), the average temperature of the sample drops by 0.5 ± 0.1 °C.

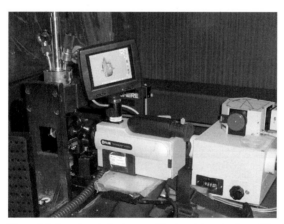

**Figure 4.5** Experimental setup used for the determination of the temperature field of the sample.

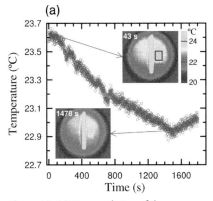 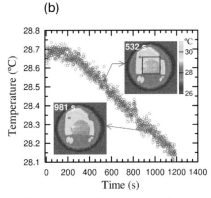

**Figure 4.6** (a) Time evolution of the average temperature of the Er$^{3+}$:KPb$_2$Cl$_5$ at 870 nm. The insets show colormaps of the temperature field of the whole system (sample plus cryostat) at two different times as measured with the thermal camera. The *rectangle* in the *upper inset* delimits the area used to calculate the average temperature of the sample. (b) Same as (a) for the Er$^{3+}$:CNBZn sample. The laser wavelength was 860 nm in this case.

Estimates for the expected bulk temperature change can be obtained from first principles using the boson operator elimination method developed by Petrushkin, Samartev and Adrianov [25, 26]. According to these authors, the macroscopic steady-state temperature change in the high-temperature limit is given by

$$\Delta T \approx -\frac{\hbar \omega_0}{k_B} \frac{\tau_v N}{\tau M}, \tag{4.7}$$

where $\tau_v$ is the mean lifetime of the vibrational modes of the host material, $\tau$ is the radiative lifetime of the optically active ion excited state, $N$ is the number density of optically active ions in the host material, and $M$ is the number density of phonon modes participating in the cooling process. For the crystal sample, taking the values $\tau = 2.4$ ms [27], $\tau_v \approx 10^{-6}$ s, $M \approx 10^{16}$ cm$^{-3}$ [25], and $N = 5.2 \times 10^{19}$ cm$^{-3}$ [27], one gets $\Delta T \approx -6.3\,°C$. On the other hand, for the glass sample, taking the values $\tau = 1$ ms, $\tau_v \approx 10^{-6}$ s, $M \approx 2.5 \times 10^{16}$ cm$^{-3}$, and $N = 10^{20}$ cm$^{-3}$, one gets $\Delta T \approx -11.68\,°C$. The discrepancy between the theoretical estimates and the present experimental results can be attributed to partial reabsorption of the anti-Stokes fluorescence (which is not taken into account in these models) or additional absorption processes of the pumping radiation involving excited states (which are known to be significant in these materials) [27]. In any case, these small temperature changes should not come as a surprise if one takes into account the minute concentrations of the optically active ions in the materials studied in this work, the geometry of the cooling experiment (single-pass configuration), and the relatively short duration of the experimental runs. On the contrary, we think that the fact that bulk cooling is obtained in spite of these difficulties indicates that CASE in these materials is extremely efficient.

## 4.4
### Influence of Upconversion Processes on the Cooling Efficiency of $Er^{3+}$

#### 4.4.1
#### Spectroscopic Grounds: Upconversion Properties of $Er^{3+}$ Under Pumping in the $^4I_{9/2}$ Manifold

The results of the previous section show that cooling was obtained by exciting the $Er^{3+}$ ions at the low-energy side of the $^4I_{9/2}$ manifold with a tunable Ti:sapphire laser. This excited state where cooling can be induced is also involved in infrared-to-visible upconversion processes. It is worth mentioning that excitation at the $^4I_{9/2}$ level is followed by fast nonradiative decay to the $^4I_{11/2}$ level in oxide and fluoride systems, due to their relatively high phonon energies. Thus, level $^4I_{9/2}$ is not available as a starting level for upconversion. However, in these low phonon energy materials, due to the reduction of multiphonon relaxation rates and the increased lifetime of the $^4I_{9/2}$ level, this one acts as an intermediate state for upconversion processes. Upconversion can occur by radiative excited state absorption (ESA) and by nonradiative energy transfer upconversion between two excited ions [28, 29]. Excited state absorption occurs when infrared photons are sequentially absorbed within a single ion, raising its energy to a higher excited level. The energy transfer upconversion process occurs when two neighboring ions in an excited state interact and one ion transfers its energy to the other one, which is promoted to a higher excited level. In this section we present the upconversion processes following excitation at the $^4I_{9/2}$ level in the low phonon energy materials under study. In the case of potassium lead chloride crystal, resonant excitation of the $^4I_{9/2}$ level leads to fluorescence from the $^2H_{9/2}$ and $^4S_{3/2}$ levels, whereas excitation at lower energies than the one of level $^4I_{9/2}$ results in emission mainly from level $^2H_{9/2}$. This latter upconversion emission can be attributed to a sequential two-photon absorption. However, under resonant pumping of the $^4I_{9/2}$ level, the mechanisms that populate the $^2H_{9/2}$ and $^4S_{3/2}$ levels are energy transfer upconversion processes.

Visible upconversion has been observed at room temperature in both samples under continuous wave (cw) laser excitation at the $^4I_{9/2}$ level. The upconverted emission spectra were measured using a Ti:sapphire ring laser. Cut-off filters were used to remove the pump radiation. As an example, Figure 4.7 shows the room-temperature upconverted emission spectra of $Er^{3+}$-doped $KPb_2Cl_5$ crystal obtained under cw near-infrared excitation at 801 nm in resonance with the $^4I_{15/2} \rightarrow {}^4I_{9/2}$ transition and at 821 nm. The observed emissions under excitation at 801 nm correspond to transitions $^2H_{9/2} \rightarrow {}^4I_{15/2}$, $^2H_{11/2} \rightarrow {}^4I_{15/2}$, $^4S_{3/2} \rightarrow {}^4I_{15/2}$, $^2H_{9/2} \rightarrow {}^4I_{13/2}$, and $^2H_{9/2} \rightarrow {}^4I_{11/2}$. The $^2H_{11/2} \rightarrow {}^4I_{15/2}$ transition is only observed at room temperature because the $^2H_{11/2}$ level is populated from $^4S_{3/2}$ via a fast thermal equilibrium between levels. As can be observed, the most intense emission corresponds to the green one from level $^4S_{3/2}$. Upon excitation at 821 nm, below the $^4I_{9/2}$ level, the upconverted luminescence spectrum is mainly characterized by the emission from level $^2H_{9/2}$. The spectra show three main bands corresponding to transitions $^2H_{9/2} \rightarrow {}^4I_{15/2,13/2,11/2}$.

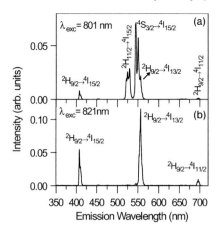

Figure 4.7 Room-temperature upconversion emission spectra obtained at two different excitation wavelengths (a) 801 nm and (b) 821 nm for KPb$_2$Cl$_5$ crystal.

The dependence of the upconverted emission intensity on the pump power is quadratic, which indicates a two-photon upconversion process to populate the $^2H_{9/2}$ and $^4S_{3/2}$ levels. This in turn may be associated with excited state absorption (ESA) and/or energy transfer upconversion (ETU). The analysis of the experimental results as a function of excitation wavelength and time confirms the existence of two different upconversion mechanisms (ESA and ETU) to populate level $^2H_{9/2}$ depending on the excitation energy.

Excitation spectra of the upconverted luminescence provide a first option for distinguishing between ESA and ETU mechanisms. In the case of ETU, the excitation spectrum is proportional to the square of the ground-state absorption coefficient as a function of wavelength, whereas the upconversion excitation spectrum is the result of the ground-state absorption and the excited-state absorption in the case of excited-state absorption. In agreement with these arguments, the excitation spectrum of the upconverted luminescence from level $^2H_{9/2}$ shows the presence of peaks corresponding to the $^4I_{15/2} \rightarrow {}^4I_{9/2}$ absorption of Er$^{3+}$ ions, which clearly indicates that an ETU process from the $^4I_{9/2}$ multiplet takes place. However, the spectrum also presents additional intense peaks at energies that are resonant with the $^4I_{9/2} \rightarrow {}^2H_{9/2}$ transition. This indicates that ESA is the mechanism responsible for the $^2H_{9/2}$ emission obtained under nonresonant excitation at the ESA peaks in the excitation spectrum [27]. Figure 4.8 shows an example of the excitation spectrum of the visible luminescence from the $^2H_{9/2}$ level. The excitation spectrum of the upconverted emission from the $^4S_{3/2}$ level follows the same wavelength dependence as the one-photon absorption spectrum, which indicates that we are dealing with an energy transfer upconversion (ETU) to populate the $^4S_{3/2}$ state.

The upconverted emission spectra in the glass sample also show a bright green emission corresponding to the $^4S_{3/2} \rightarrow {}^4I_{15/2}$ and $^2H_{9/2} \rightarrow {}^4I_{13/2}$ transitions, together with weak UV and visible (blue and red) emissions (see Figure 4.9). In this case we observed the green emission from both levels for all excitation wavelengths.

The excitation spectrum of the upconverted luminescence resembles the one-photon absorption spectrum at high energies but is broader and presents some peaks superimposed on the ground state absorption, which suggests the presence

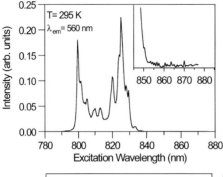

Figure 4.8 Excitation spectrum of the upconverted emission at 560 nm of $Er^{3+}$ in $KPb_2Cl_5$ crystal.

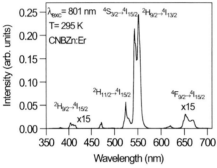

Figure 4.9 Room-temperature upconversion emission spectra obtained at 801 nm for CNBZn glass.

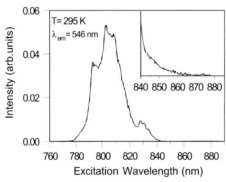

Figure 4.10 Excitation spectrum of the upconverted emission at 546 nm of $Er^{3+}$ in CNBZn glass.

of ETU and ESA processes. In this case, due to the inhomogeneous broadening it is difficult to select only the emission from the $^2H_{9/2}$ level. Figure 4.10 shows the excitation spectrum obtained by collecting the luminescence of the $^4S_{3/2}$ level at 546 nm.

We measured the upconversion emission under excitation at wavelengths corresponding to the cooling region for both samples. In this case, we observed emission from the $^4S_{3/2}$ and $^2H_{9/2}$ levels. As an example, Figure 4.11 shows the green upconverted emissions of $Er^{3+}$ from both samples obtained under infrared excitation at 854 and 856 nm for the crystal and glass respectively. The observed emissions correspond to transitions $^2H_{11/2} \rightarrow {}^4I_{15/2}$, $^4S_{3/2} \rightarrow {}^4I_{15/2}$, and $^2H_{9/2} \rightarrow {}^4I_{13/2}$. The $^2H_{11/2}$ level is populated from the $^4S_{3/2}$ level via a fast thermalization between both levels.

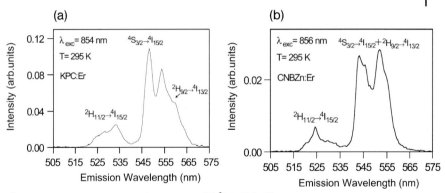

**Figure 4.11** Upconversion emission spectra of Er$^{3+}$ in KPb$_2$Cl$_5$ (KPC) crystal and CNBZn glass obtained under excitation in the cooling regions.

### 4.4.2
### A Phenomenological Cooling Model Including Upconversion

As pointed in the previous section and reported by the present authors previously [30], energy upconversion plays an important role in these systems precisely in the spectral region where laser cooling takes place. One could immediately ask whether the additional upconversion channels can contribute to the laser cooling process. Intuitively, one would expect that these additional channels would enhance the cooling efficiency, due to the fact that photons with much larger energy than the one of the absorbed photons are emitted in each instance of these processes, as has been pointed out in the past by other authors [31, 32].

To place these observations into a more formal context, we have devised a simple phenomenological model that allows one to calculate the cooling efficiency under steady-state conditions. This model has its starting point at the energy level diagram of Figure 4.12 and the absorptions and emissions depicted there. Accordingly, the rate equations that describe the population dynamics of this system are

$$\frac{dN_1}{dt} = -W_{10}N_1 + \beta^{(2)}W_{21}N_2 + \gamma N_2^2 + \beta^{(3)}W_{31}N_3 + \beta^{(4)}W_{41}N_4 + \left(1-\eta_e^{(1)}\right)$$
$$\times W_{rad}^{(1)}N_1, \tag{4.8a}$$

$$\frac{dN_2}{dt} = \frac{P_{abs}^r}{h\nu} - \alpha^{(2)}W_{20}N_2 - \beta^{(2)}W_{21}N_2 - 2\gamma N_2^2 - \sigma(\nu)\frac{I_p(t)}{h\nu}N_2 + \left(1-\eta_e^{(2)}\right)$$
$$\times W_{rad}^{(2)}N_2, \tag{4.8b}$$

$$\frac{dN_3}{dt} = \gamma N_2^2 - \alpha^{(3)}W_{30}N_3 - \beta^{(3)}W_{31}N_3 + \left(1-\eta_e^{(3)}\right)W_{rad}^{(3)}N_3, \tag{4.8c}$$

$$\frac{dN_4}{dt} = \sigma(\nu)\frac{I_p(t)}{h\nu}N_2 - \alpha^{(4)}W_{40}N_4 - \beta^{(4)}W_{41}N_4 + \left(1-\eta_e^{(4)}\right)W_{rad}^{(4)}N_4 \tag{4.8d}$$

and the conservation of population relation $N_0 + N_1 + N_2 + N_3 + N_4 = N$. In these expressions, $P_{abs}^r$ stands for the resonantly absorbed power, $\sigma(\nu)$ is the excited-state

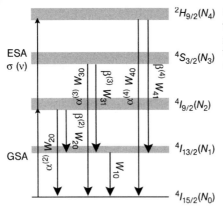

**Figure 4.12** The energy levels involved in the upconversion processes. GSA stands for ground-state absorption whereas ESA stands for excited state absorption. See text for the meanings of the *other symbols*.

absorption cross-section at pumping frequency $\nu$, $I_p(t)$ is the intensity of the incident beam, $W_{\text{rad}}^{(i)}$ is the radiative decay rate of the $i$-th level, $W_{ij} = W_{\text{rad}}^{(i)} + W_{\text{nr}}^{ij}$ is the sum of the $i$-th level radiative decay rate, $W_{\text{rad}}^{(i)}$, and the nonradiative decay rate from that level to level $j$, $W_{\text{nr}}^{ij}$. The last term in the latter two equations describes repopulation by reabsorption of the fluorescence for the considered level. On the other hand, $\alpha^{(i)}$ and $\beta^{(i)}$ are the branching ratios for the $i \to 0$ and $i \to 1$ transitions, respectively, and they verify $\alpha^{(i)} + \beta^{(i)} = 1$. $\gamma N_2^2$ describes an energy transfer upconversion (ETU) process in which two $Er^{3+}$ ions at the $^4I_{9/2}$ level interact, with one ion gaining energy and reaching the $^4S_{3/2}$ level and the other one losing energy and moving to the $^4I_{13/2}$ level. This ETU was found to be responsible for the long-time behavior of the $^4S_{3/2}$ dynamics in $Er^{3+}$:$KPb_2Cl_5$ by the present authors [27].

Using (4.8) and following the approach developed by Hoyt and coworkers [32], one can easily calculate the cooling efficiency to leading order in the ESA and ETU processes after a little algebra:

$$\eta_{\text{cool}} = \alpha_b + \alpha_r(\nu) \left\{ 1 - \tilde{\eta}_q^{(2)} \frac{\bar{\nu}_f^{(2)}}{\nu} - \gamma \tilde{\eta}_q^{(3)} \frac{\bar{\nu}_f^{(3)}}{\nu} \left( \frac{\tilde{\eta}_q^{(2)} \tau_{\text{rad}}^{(2)}}{\eta_e^{(2)}} \right)^2 \frac{\alpha_r}{\alpha_T} \frac{P_{\text{abs}}}{h\nu} \right.$$
$$\left. - \tilde{\eta}_q^{(4)} \frac{\tilde{\eta}_q^{(2)} \tau_{\text{rad}}^{(2)}}{\eta_e^{(2)}} \frac{\sigma(\nu)}{A} \frac{\bar{\nu}_f^{(4)}}{\nu} \frac{P_{\text{in}}}{h\nu} \right\}, \qquad (4.9)$$

where $\alpha_b$ is a background absorption coefficient that is nearly independent of frequency, $\alpha_r$ is the resonant part of the absorption coefficient, $P_{\text{abs}} = P_{\text{abs}}^r + P_{\text{abs}}^b$, and $P_{\text{in}}$ is the input laser power; we have defined the generalized quantum efficiency for level $i$

$$\tilde{\eta}_q^{(i)} = \frac{\eta_e W_{\text{rad}}^{(i)}}{\eta_e W_{\text{rad}}^{(i)} + \alpha^{(i)} W_{\text{nr}}^{i0} + \beta^{(i)} W_{\text{nr}}^{i1}}, \qquad (4.10)$$

and the weighted fluorescence frequency from level $i$

$$\bar{\nu}_f^{(i)} = \alpha^{(i)} \nu_f^{i0} + \beta^{(i)} \nu_f^{i1}. \qquad (4.11)$$

We have also used the relation $W_{\text{rad}}^{(i)} = (\tau_{\text{rad}}^{(i)})^{-1}$. In order to simplify the discussion below, let us also consider the limits $\tilde{\eta}_q^{(i)} \approx 1$, $\eta_e^{(i)} \approx 1$, and $\alpha_T L \ll 1$, where $L$ is the length traveled by the laser beam inside the sample, so that $P_{\text{abs}} \approx P_{\text{in}} \alpha_T L$ (low pump-depletion limit). With these simplifications, the cooling efficiency reads

$$\eta_{\text{cool}} = a_b + a_r(\nu) \left\{ 1 - \frac{\bar{\nu}_f^{(2)}}{\nu} - \gamma \frac{\bar{\nu}_f^{(3)}}{\nu} \left( \tau_{\text{rad}}^{(2)} \right)^2 a_r(\nu) \frac{P_{\text{in}} L}{h\nu} - \tau_{\text{rad}}^{(2)} \frac{\sigma(\nu)}{A} \frac{\bar{\nu}_f^{(4)}}{\nu} \frac{P_{\text{in}}}{h\nu} \right\}. \tag{4.12}$$

The condition for cooling to occur is given by

$$\eta_{\text{cool}} \le 0. \tag{4.13}$$

An interesting special case is pure anti-Stokes cooling ($\gamma = \sigma = 0$). Neglecting the $\nu$ dependence of $a_r(\nu)$ (around the center of a Gaussian-shaped absorption peak, for example), the condition for cooling is given by

$$\bar{\nu}_0 = \frac{a_r}{a_T} \bar{\nu}_f \tag{4.14}$$

or, equivalently

$$\bar{\lambda}_0 = \frac{a_T}{a_r} \bar{\lambda}_f, \tag{4.15}$$

where $\bar{\lambda}_f = \frac{c}{\bar{\nu}_f}$ is the average fluorescence wavelength. To quote an example, for Er$^{3+}$:KPb$_2$Cl$_5$, $\bar{\lambda}_0 = 957$ nm if one considers all of the emissions from the $^4I_{9/2}$ level. In the general case, the exact frequency cutoff depends on both $a_r(\nu)$ and $\sigma(\nu)$. However, a general conclusion that one can extract from (4.12) is that the onset of cooling will occur at higher frequencies with regards to $\bar{\nu}_0$ in the presence of the upconversion channels, as expected. To make a rough estimate of the cooling cutoff shift due to upconversion, let us consider the limit in which both $a_r(\nu) = a_0$ and $\sigma(\nu) = \sigma_0$ are frequency independent. For small values of both $\gamma$ and $\sigma_0$, one finds that the onset of cooling occurs at

$$\bar{\nu}_0' = \bar{\nu}_0 + 2 \frac{\nu_{\text{up}}^2}{\bar{\nu}_0}, \tag{4.16}$$

where

$$\nu_{\text{up}}^2 = \frac{a_r}{a_T} \frac{\gamma h \bar{\nu}_f^{(3)} \left( \tau_{\text{rad}}^{(2)} \right)^2 a_0 P_{\text{in}} L - \tau_{\text{rad}}^{(2)} \frac{\sigma_0}{A} h \bar{\nu}_f^{(4)} P_{\text{in}}}{h^2},$$

which is always larger than $\bar{\nu}_0$. Taking typical values for the various parameters in (4.16): $\gamma \approx 0.5 \times 10^{-9}$ s$^{-1}$, $\tau_{\text{rad}}^{(2)} = 2.4$ ms [27], $a_0 \approx 10^{-4}$ m$^{-1}$, $\bar{\nu}_f^{(3)} = 16\,382$ cm$^{-1}$, $\bar{\nu}_f^{(4)} = 17\,953$ cm$^{-1}$, $P_{\text{in}} = 2$ W, $L \approx 10$ mm, $A \approx 1$ mm$^2$, and $\sigma_0 \approx 10^{-21}$ cm$^2$, one gets $\bar{\lambda}_0' = 874$ nm for Er$^{3+}$:KPb$_2$Cl$_5$.

Another interesting feature predicted by this model is the possibility of obtaining a net cooling effect, even above the frequency cutoff, by increasing the pumping power. To see this, let us consider the situation $\nu > \bar{\nu}_0$ in the limit $\alpha_b \ll \alpha_r$. Condition (4.13) in this case leads to the existence of a power threshold given by

$$P_{in}^0 = \frac{(\nu - \bar{\nu}_0)h\nu}{\gamma h\bar{\nu}_f^{(3)} \left(\tau_{rad}^{(2)}\right)^2 \alpha_r L + \tau_{rad}^{(2)} \frac{\sigma}{A} h\bar{\nu}_f^{(4)}} \quad (4.17)$$

above which cooling will occur.

Therefore, as expected, the cooling efficiency of a material where efficient upconversion processes are present is enhanced with respect to standard materials due to the additional cooling channels provided by the upconversion mechanisms that operate.

## References

1 KUMTA, P.N. AND RISBUD, S.H. (1994) *J. Mater. Sci.*, **29**, 1135.

2 PAGE, R., SHAFFERS, K., PAYNE, S. AND KRUPKE, W. (1997) *J. Lightwave Tech.*, **15**, 786.

3 NIKL, M., NITSCH, K., AND POLÁK, K. (1991) *Phys. Stat. Sol. B* **166**, 511.

4 NIKL, M., NITSCH, K., VELICKÁ, I., HYBLER, J., POLÁK, K. AND FABIAN, T. (1991) *Phys. Stat. Sol. B*, **168**, K37.

5 NITSCH, K., DUŠEK, M., NIKL, M., POLÁK, K. AND RODOVÁ, M. (1995) *Prog. Cryst. Growth Ch.*, **30**, 1.

6 WEBER, M.J. (1982) *J. Non-Cryst. Solids*, **47**, 117.

7 LUCAS, J. AND ADAM, J.L. (1989) *Glastech. Ber.*, **62**, 422 and references therein.

8 VAN UITER, L.G. AND WEMPLE, S.H. (1978) *Appl. Phys. Lett.*, **33**, 57.

9 ANGELL, C.A., CHANGH, L. AND SUNDAR, H.G.K. (1985) *Mat. Sci. Forum*, **5**, 189.

10 MATECKI, M., POULAIN, M. AND POULAIN, M. (1987) *Mat. Sci. Forum*, **19/20**, 47.

11 Adam, J.L., Matecki, M., H. L'Helgoualch, and Jaquier, B. (1994) *Eur. J. Solid State Inorg. Chem.*, **31**, 337.

12 ISAENKO, L., YELISSEYEV, A., TKACHUK, A., IVANOVA, S., VATNIK, S., MERKULOV, A., PAYNE, S., PAGE, R. AND NOSTRAND, M. (2001) *Mater. Sci. Eng. B*, **81**, 188.

13 CASCALES, C., FERNÁNDEZ, J. AND BALDA, R. (2005) *Opt. Exp.*, **13**, 2141.

14 GABRIEL, A. AND PELTON, A.D. (1985) *Can. J. Chem.*, **63**, 3276.

15 SUMAROKOVA, T.N. AND MODESTOVA, T.P. (1960) *Zh. Neorg. Khim.*, **6**, 679.

16 ECKSTEIN, J., NITSCHE, R., TRAUTH, J. AND GUTMAN, R. (1988) *Mat. Res. Bull.*, **23**, 813.

17 WILLEMSEN, B. (1971) *J. Solid State Chem.*, **3**, 567.

18 LÉBL, M. AND TRNKA, J. (1965) *Z. Physik*, **186**, 128.

19 Singh, N.B., Duval, W.M.B. and Rosenthal, B.N. (1988) *J. Cryst. Growth*, **89**, 80.

20 URSU, I., NISTOR, S.V., VODA, M.M., NISTOR, L.C. AND TEODORESCU, V. (1993) *Mat. Res. Bull.*, **18**, 1275.

21 VODA, M. AND TRUTIA, A. (1969) *Rev. Roum. Phys.*, **14**, 551.

22 NOSTRAND, M.C., PAGE, R.H., PAYNE, S.A., KRUPKE, W.F., SCHUNEMAN, P.G. AND ISAENKO, L.I. (1998) *OSA TOPS*, **19**, 524.

23 FERNÁNDEZ, J., GARCÍA-ADEVA, A.J. AND BALDA, R. (2006) *Phys. Rev. Lett.*, **97**, 033001.

24 JACKSON, W.B., AMER, N.M., BOCCARA, A.C. AND FOURNIER, D. (1981) *Appl. Opt.*, **20**, 1333.

25 PETRUSHKIN, S.V. AND SAMARTSEV, V.V. (2001) *Theor. Math. Phys.*, **126**, 136.

26 Andrianov, S.N. and Samartsev, V.V. (1999) *Laser Phys.*, **9**, 1021.
27 Balda, R., Garcia-Adeva, A.J., Voda, M. and Fernandez, J. (2004) *Phys. Rev. B*, **69**, 205203.
28 Auzel, F. (1973) *Proc. IEEE*, **61**, 758.
29 Wright, J.C. (1976) In *Radiationless processes in molecules and condensed phases*, (ed F.K. Fong), Heidelberg, Springer-Verlag, 239–295.
30 Garcia-Adeva, A.J., Balda, R. and Fernandez, J. (2007) *Proc. SPIE*, **6461**, 646102.
31 Hoyt, C.W., Hasselbeck, M.P., Sheik-Bahae, M., Epstein, R.I., Greenfield, S., Thiede, J., Distel, J. and Valencia, J. (2003) *J. Opt. Soc. Am. B*, **20**, 1066.
32 Hoyt, C.W., Sheik-Bahae, M., Epstein, R.I., Edwards, B.C. and Anderson, J.E. (2000) *Phys. Rev. Lett.*, **85**, 3600.

# 5
# Laser Refrigerator Design and Applications
*Gary Mills and Mel Buchwald*

## 5.1
## Introduction

The simplest implementation of a refrigerator based on the principle of anti-Stokes fluorescence is a ytterbium Yb:ZBLAN cylinder (cooling element) with high-reflectivity dielectric mirrors deposited on the ends, as shown in Figure 5.1. A photo of such a cooling element is shown in Figure 5.2. The pump beam is introduced through a small hole in one mirror and bounces back and forth until it is absorbed or escapes. A key feature of this arrangement is that the pump light is confined to a nearly parallel beam while the fluorescence is emitted randomly and into $4\pi$ steradians. This makes it possible to allow the fluorescence to escape while the pump light is trapped. The fluorescent photons that are nearly parallel to the pump beam do not escape. They are reabsorbed and ultimately escape with a small and calculable degradation of the overall efficiency.

Figure 5.3 shows an early design concept for an optical cryocooler detector dewar. The cooling element is bonded directly to the focal plane structure in order to absorb the heat. The fluorescence is absorbed by a heat sink with high absorbtivity. The cooling element and focal plane are supported by a folded tube of low thermal conductivity material, in a manner similar to dewars used for focal planes

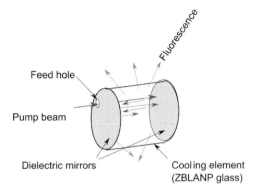

**Figure 5.1** Dielectric mirrors provide a long pump path.

*Optical Refrigeration. Science and Applications of Laser Cooling of Solids.*
Edited by Richard Epstein and Mansoor Sheik-Bahae
Copyright © 2009 WILEY-VCH Verlag GmbH & Co. KGaA, Weinheim
ISBN: 978-3-527-40876-4

**Figure 5.2** A Yb:ZBLAN fluorescent element provides the cooling in prototype optical refrigerators. The dielectric mirror coatings are highly reflective at 1035 nm but transparent at visible wavelengths.

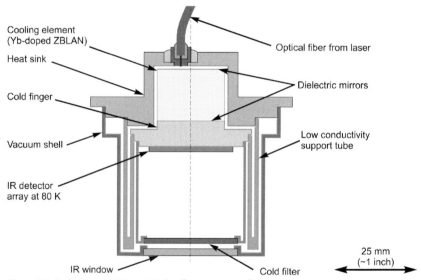

**Figure 5.3** Early (circa 2000) anti-Stokes fluorescence refrigerator design concept for an infrared detector.

cooled by mechanical cryocoolers. Note that the cooling element with its heat sink is small compared to the structure needed to support a focal plane at cryogenic temperatures. The entire refrigerator is contained in a small vacuum chamber, which contains an IR window for the detector and an optical fiber to bring the pumping light from the laser.

A copper heat sink completely surrounds the cooling assembly and is mounted to the vacuum chamber wall. A significant issue with the heat sink is the properties of the surface facing the cooling assembly. This surface needs to selectively absorb the near 1 μm fluorescence while exhibiting low emittance to the ambient black-body radiation. The assembly, which includes the cooling element, thermal link and the load mass, is mechanically supported within the heat sink using a fiberglass-epoxy support.

The cooling element is made from a Yb:doped zirconium fluoride glass (ZBLAN). It is cylindrical and coated on both ends with high-performance dielectric mirrors. It is typically doped with 2% (by mass) ytterbium fluoride.

The design shown in Figure 5.3 has most of the features of a feasible laser cryocooler, but assumes very small leakage of light through the dielectric mirrors. The failure of this assumption was not discovered until laboratory observations of this design were performed. This problem and its resolution are discussed below in the section on design issues. Modifications were required to produce a working cryocooler. This conceptual design, though, is a useful reference point for discussions of cryocooler modeling.

## 5.2 Modeling

The performance of an optical cryocooler is limited by two types of effects: escape of pump photons from the cooling element before they can cause cooling, and absorption of pump or fluorescent photons by processes that produce heat. The heating effects are by far the more serious.

The known pump photon loss mechanisms are: escape through the mirrors, escape past the edges of the mirrors, escape through the feed hole, and scattering in the bulk material. The known heating mechanisms are direct absorption by contaminants; absorption by $Yb^{3+}$, followed by "hopping" of the excitation to a contaminant (especially Fe); absorption in the mirrors or by material beneath the mirrors; absorption by the cold stage of the cryocooler of photons not absorbed by the heat sink; and radiative and conductive heat load from the cryocooler structure.

Three additional effects come into play: saturation of the $Yb^{3+}$ absorption at high-power densities, and reabsorption and recycling of those fluorescent photons that cannot escape from the sides of the cooling element. Readsorption of the bluest part of the fluorescence leads to a reddening of the fluorescence that escapes. These effects are calculable and are not a significant limitation in an optimized design. Because these many effects interact with the design parameters in different ways, a comprehensive simulation is required to predict the performance and to understand the design tradeoffs involved.

Our group at Ball has developed a comprehensive numerical model for optical refrigerators. It is a valuable design tool that provides a way of rapidly exploring and optimizing the design parameters, and permits evaluation of the refrigerator for specific cooling applications. The model consists of three parts: tracing the life history of the incoming pump photons, tracing the life history of the outgoing fluorescent photons, and evaluating the internal and external heat transfer.

The first part of the model calculates the probabilities of five possible fates for each pump photon:

1. Absorption by Yb, which includes the reduction by saturation at high power densities
2. Absorption by anything else, including unknown contaminants that can only be described empirically
3. Leakage from mirrors
4. Leakage through feed hole
5. Leakage at mirror edges

The absorption and emission of photons is a function of temperature, Yb doping, and optical intensity. It is calculated based on a physical model of the temperature-dependent population distributions, which includes detailed $Yb^{3+}$ ion energy level distributions and the transition strengths between ion levels. The adjustable parameters of this physical model were obtained from fitting Yb:ZLAN absorption and emission data obtained at several temperatures. We do not consider laser pump powers where Boltzmann statistics are not suitable.

The second part of the model calculates the four possible fates of the emitted photons:

1. Escape through the ends (mirrors)
2. Escape through the sides
3. Reabsorption by Yb (recycling)
4. Absorption by anything else, which is assumed to cause heating

The number of fluorescent photons that escape or are reflected back into the cooling element is a function of the refractive index of the Yb:ZBLAN and the element geometry. We assume that the excitation of the cylinder volume is uniform. The average energy of the photons escaping the cooling element is calculated, including the effect of "reddening". The ultimate output of the model is the cooling efficiency for the fluorescent process, calculated by subtracting the pump light energy from the escaping fluorescent energy, and dividing by the pump energy.

The third part of the model calculates the conductive and radiative parasitic head loads from the refrigerator structure, plus the temperature drop within the cooling element, using standard heat transfer analysis techniques. Yb:ZBLAN exhibits high absorption of the 10 µm thermal radiation but is surrounded by a close-fitting low-emittance heat sink. The parallel-plate gray surface approximation is used for radiation view factors.

The most significant limitation of the model at this time is its restriction to Yb:ZBLAN. We have detailed temperature-dependent emission and absorption data for this cooling material alone. In addition, the model assumes isotropic absorption and emission, which is valid for ZBLAN and other glassy hosts, but would not in general be valid for crystal hosts that can have different optical properties for each crystal axis.

## 5.3 Modeling Results

The model was used to analyze the operation of a small optical cryocooler such as that shown in Figure 5.3. The cooling element is a ZBLAN-glass cylinder doped with 2% Yb which is 15 mm long with a cross-sectional area of 50 mm$^2$. The ends are coated with dielectric mirrors with 99.999% reflectance (which has already been achieved on Yb:ZBLAN). The cooling efficiency was compared at high pump power (4 W) and low pump power (10 mW) across a range of temperatures and pump wavelengths.

Figure 5.4 shows that the cooling efficiency drops with decreasing temperature, as expected, because of the reduced populations in the upper energy levels caused by significant cooling.

For each temperature there is an optimum wavelength that results in maximum cooling efficiency, due to competing physical phenomena. Increasing the pump wavelength tends to increase the efficiency by increasing the energy difference between the pump photon and the average fluorescent photon (approximately 990 nm). At longer wavelengths and lower temperatures the absorption length that the pump light must travel increases, reducing the efficiency because of mirror leakage and other losses that grow in importance with absorption path length. At

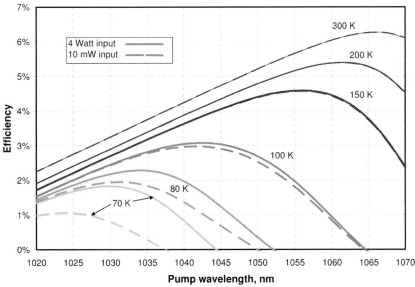

**Figure 5.4** Theoretical cooling element performance of a 15 mm long Yb:ZBLAN cooling element as a function of temperature and wavelength.

the lowest temperatures the effects of saturation of the absorption at high pump power become appreciable.

Figure 5.5 shows the available heat lift calculated for this design. The parasitic heat leak to the cold stage is independent of the pump power. The total heat lift at the optimum wavelength varies with temperature and input power as shown. The difference between the heat lift and the parasitic heat load is the cooling power available for a payload. The figure shows that, for Yb:ZBLAN, the practical temperature limit is around 70 K.

Our model was used to determine the suitability of an optical refrigerator for cooling a small, portable, terahertz detector in a microcryostat such as that discussed in the section on applications. The heat leak into the 90 K cold stage of the cryostat was estimated to be 5 mW.

The cooling performance was modeled to determine the configuration that would most efficiently refrigerate this load. YbF doping of 2% in the ZBLAN and 99.999% mirror reflectance was assumed. Preliminary modeling showed that the performance was very sensitive to the feed hole size, so a minimum practical diameter of 50 μm was assumed. Figure 5.4 shows that the most efficient wavelength for 90 K is 1035 nm, and that was chosen. Cooling element lengths of 2.5 to 40 mm were compatible with the cryostat design and were modeled with various cross-sectional areas. Figure 5.6 shows that the power required and the efficiency of the cooling element improve with length as the pump light absorption coefficient is reduced. Increasing the cross-sectional area initially increases the heat lift, as it reduces the saturation level. However, where the saturation level is already low, increasing the cross-section reduces the heat lift due to reabsorption of fluorescence.

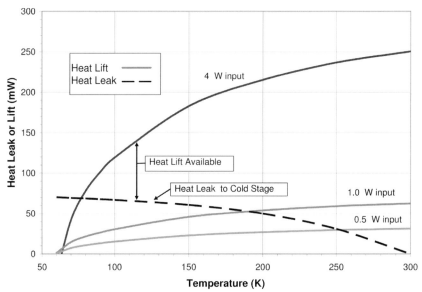

**Figure 5.5** 15 mm long Yb:ZBLAN cooling element performance at optimum pump wavelength as a function of temperature.

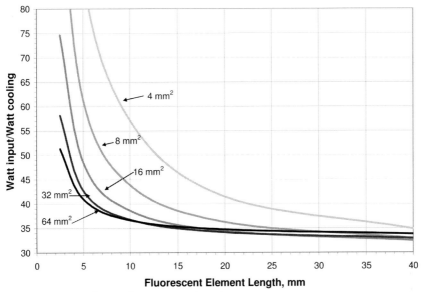

**Figure 5.6** Power efficiency for various cooling element configurations pumped with 300 mW at 1035 nm with 2% Yb doping.

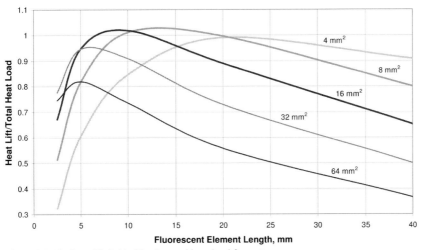

**Figure 5.7** The heat lift divided by the total heat load for various cooling element configurations with a payload heat leak of 5 mW at 90 K with an input power of 300 mW.

The reabsorption causes a reddening of the escaped fluorescence, moving it closer to the pump wavelength.

The cooling efficiency drops off rapidly when the cooling element length is less than 5 mm. This appears to be the lower limit of the practical size of a Yb:ZBLAN fluorescent element. It is clear from Figure 5.6 that increasing the length of the cooling element increases the cooling efficiency and heat lift. However, increasing the length and cross-sectional area also increases the external surface area of the cooling element and the amount it contributes to the total cold stage heat leak. To account for this we calculated the total heat load (5 mW fixed plus the cooling element radiative heat load) and divided it into the heat lift shown in Figure 5.6. The result is shown in Figure 5.7. The optimum fluorescent length appears to be around 10 mm, with an optimum cross-section of between 8 and 16 mm$^2$, or dimensions perpendicular to the length of between 2.8 and 4 mm. This shows the utility of a comprehensive model in optimizing the optical refrigerator design.

## 5.4
### Design Issues

The first tests of laboratory optical refrigerators ended in failure: when pumped with the laser light at a wavelength that should have resulted in cooling, they heated. This was in spite of the fact that the same Yb:ZBLAN material had been shown to cool with both the photothermal deflection technique and with thermal camera measurements [1, 2].

The design of the refrigerators was similar to that shown in Figure 5.3. Leakage of light through the dielectric mirrors at large angles of incidence had been observed and was suspected to be a principal cause. To understand this in more detail, we analyzed the performance of mirror stacks as a function of incident angle, using commercially available thin-film software [3–5].

Dielectric mirrors are most commonly used (e.g. laser cavities) with the incident light on the vacuum or air side and with near-normal angles of incidence. In our case, the performance of the mirror is quite high until about 50 degrees off perpendicular, where the reflectivity falls off for the *P*-polarization. The *S*-polarization is not affected.

The situation is different again when the light is incident on the mirror from the high-index substrate side. Figure 5.8 shows the result for a dielectric mirror in the arrangement required for optical cooling, where the incident medium is the Yb:ZBLAN glass cooling element. The mirror is also in physical contact with a YAG substrate acting as the thermal conductor to the load. This results in much greater mirror leakage for both polarizations.

The fluorescence is emitted within the cooling material isotropically and will be incident on the mirror at all angles. The mirror leakage shown in Figure 5.8 and the resulting absorption of energy at the load can be large enough to negate the cooling effect.

To mitigate the mirror leakage, a silver layer was assumed on the outside of the mirror stack (before the YAG). The results are shown in Figure 5.9. An optically thick layer of silver on the exit face of the quarter-wave dielectric stack does not reflect all of the light leaked at oblique angles of incidence; instead, all of the light transmitted by the dielectric coating is absorbed by the silver layer, making the layer essentially "black". The dielectric multilayer acts as an antireflection coating on the

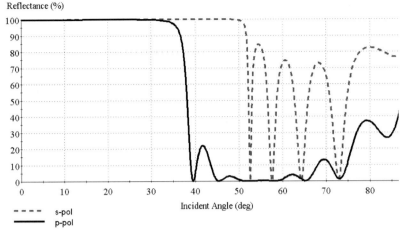

**Figure 5.8** Reflectance as a function of incident angle at 975 nm with an incident medium of ZBLAN glass. Quarter-wave stack of 27 layers of silica and tantalum pentoxide on YAG substrate at a design wavelength of 1064 nm.

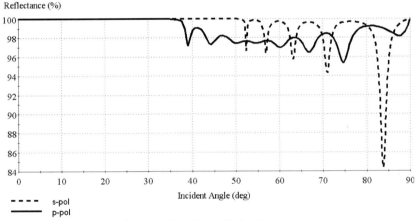

**Figure 5.9** Reflectance as a function of incident angle for the same configuration as in Figure 5.8, but with a silver layer on the YAG substrate.

silver, allowing more of the light to penetrate the silver layer where it is efficiently absorbed.

In another effort to mitigate the mirror leakage, stack materials with higher refractive index were considered. For the visible-NIR spectral region, the greatest available dielectric material refractive index contrast comes from zinc sulfide and magnesium fluoride. With dielectric stacks, a large refractive index contrast produces a large region of high reflectance as a function of incident angle. Note that even the index contrast of the ZnS and $MgF_2$ is insufficient to reflect all of the S- and P-polarized light between 40 and 90° at 975 nm, as shown in Figure 5.10. This

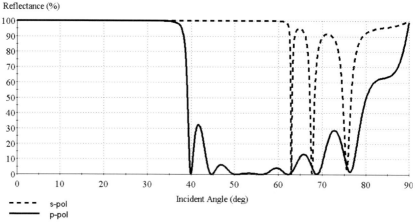

**Figure 5.10** Reflectance as a function of incident angle at 975 nm with an incident medium of Yb:ZBLAN. Quarter-wave stack of 27 layers of ZnS and MgF$_2$ on Nd:YAG substrate at 1064 nm.

behavior is typical of dielectric mirrors in the visible-NIR spectral region because of the limited choice of dielectric material in these spectral regions.

We also investigated the omnidirectional dielectric mirror designs suggested by Mansuripur [5]. We found that this approach would allow for high reflectance over broad angles of incidence at a single wavelength. However, to extend this to the wavelength range of the fluorescence cooling effect requires layer materials with very low dispersion, which are not currently available.

The effect of mirror leakage on optical cooling performance is significant. Using ray tracing software, we have calculated that 27% of the light emitted by the cooling element leaks through a typical mirror stack and might be incident on the cooling load. Since the available heat lift ranges from 1 to 6% of the fluorescence, depending on temperature, the energy from the leakage has the potential to significantly reduce or negate the cooling effect.

There appears to be no feasible way to stop the leakage of the fluorescence through the dielectric mirrors. A method is needed to keep mirror leakage from being absorbed by the cooling load. One approach is shown in Figure 5.11. The fluorescent cooling element is mounted on a thermal link made of transparent but conductive material such as sapphire or YAG. The angle on the outer edge of the bridge is chosen to optimize the escape of the fluorescent photons that leak through the dielectric mirror at angles significantly off-axis. The connection to the cooling load is sufficiently offset such that any ray leaked from the fluorescent element is reflected away from the cooling load by total internal reflection. A material with an index of refraction smaller than that of the bridge is placed between the bridge and the cooling load. This could, for example, be the adhesive that bonds the bridge to the cooling load.

The symmetrical design in Figure 5.11 is compact, which is important for practical cryocoolers, but it is inconvenient for laboratory testing of the concept. A vari-

5.4 Design Issues | 127

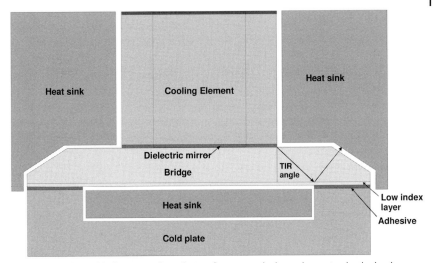

**Figure 5.11** Thermal link concept for reducing fluorescent leakage absorption by the load.

ation of this concept, more suitable for laboratory testing, is shown in Figure 5.12. For ease of manufacture, we used separate flat and angled pieces of bonded sapphire. Using an asymmetrical design with the heat sink offset to one side makes it easier to view it during tests from all angles with near-IR and thermal cameras in order to study both thermal performance and fluorescent escape. The wedge-shaped piece is designed to allow fluorescence that is otherwise trapped in the flat to escape. The design was optimized using ray tracing software [6], which predicts that less than 0.03% of the fluorescent energy is absorbed by the cooling load for an optimized design.

This prototype refrigeration assembly design has been tested. The fluorescent element was 12 mm in diameter and 13 mm long and weighed 8.5 g. The element is ZBLAN glass doped with 2% (by mass) ytterbium fluoride, with high-performance dielectric mirrors. A small hole was provided in one of the dielectric mirrors. A load mass intended to simulate a small infrared focal plane or other small sensor was bonded to the sapphire link. The load consists of an aluminum cylinder 10 mm in diameter and 6 mm thick and weighing 1.1 g. A silicon diode thermometer and an

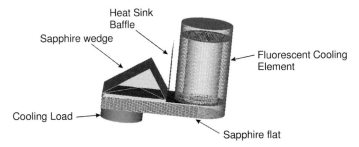

**Figure 5.12** Optimized refrigeration assembly for laboratory testing.

1/8 W, 240 Ω, carbon resistor are mounted with adhesive in a slot in the load mass cylinder. The carbon resistor was used to impose a heat load and can also serve as a secondary thermometer. Four 36-gauge phosphor bronze wires were connected to the silicon diode thermometer and carbon resistor. Photos of the assembly with no pump light and illuminated only with the 1030 nm pump beam are shown in Figure 5.13.

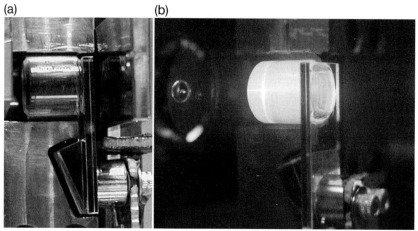

**Figure 5.13** Refrigeration assembly and heat sink with no pump and while being pumped. Images taken with a digital camera sensitive to both visible and near-infrared light. The leakage of fluorescence through the fluorescent element dielectric mirror is apparent.

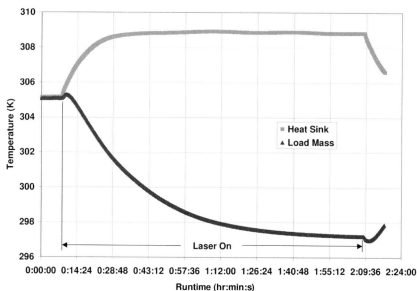

**Figure 5.14** Effect of pumping the fluorescent element with 7.4 W of 1030 nm laser light for the indicated time period.

The test refrigerator has been successfully operated [7]. With the laser beam off, the chamber containing the refrigerator was evacuated and the cooling assembly was allowed to come into thermal equilibrium with the heat sink. The laser beam was then turned on and the temperatures of the load mass and heat sink were monitored. The beam remained focused on the cooling element until a near steady state condition was achieved. The beam was then turned off and the temperatures continued to be monitored.

A typical result is shown in Figure 5.14 for 7.4 W of laser power. The fluorescence absorbed by the heat sink caused its temperature to rise until it came into thermal equilibrium with the chamber wall. At equilibrium, the load mass was cooled 7.9 °C below the starting temperature and 11.8 °C below the heat sink.

## 5.5 Mirror Heating

When the beam is introduced into the cooling element, as shown in Figure 5.14, the load mass thermometer initially heats until the cooling effect takes over and the assembly cools. When the beam is turned off, the assembly continues to cool for a brief period. We conclude from this that there is a heat source that affects the silicon diode thermometer very quickly when the beam is turned on, but that this effect is eventually overwhelmed by the cooling effect of the fluorescent element. Since the ray trace modeling showed very small amounts of fluorescence being absorbed by the cooling load, this is unlikely to be a cause of heating. A likely explanation is that the dielectric mirrors are absorbing pump light or fluorescence and are heating. Thermal images of pumped refrigeration assemblies also show the mirrors at a higher temperature. Dielectric mirror experts regard this as possible at high incident angles, but no theoretical verification has yet been suggested.

Since the dielectric mirrors also have fabrication issues when applied to ZBLAN, one possible solution is to trap the pump light in the fluorescent element with total internal reflection instead of dielectric mirrors. This concept is shown in Figure 5.15. The dielectric mirrors have been replaced with solid retroreflectors fabricated from ZBLAN. A small flat facet in the center of one of the retroreflectors provides a place for the pump beam entrance. The intent is for the pump beam to be trapped until it is absorbed, with only the fluorescence escaping.

In order to understand both the advantages and limitations of this concept, a model of the cooling element was created in existing software [6], and design parameters such as pump numerical aperture and angle of incidence were varied for Yb:ZBLAN cooling material. The goal of the modeling was to determine if this concept would be effective in trapping the pump beam or whether significant leakage would occur. Pump beam leakage has the effect of reducing the beam input power but is not as important to refrigerator performance as direct heating due to ordinary absorption.

A rendering of the modeled design is shown in Figure 5.16. The cylinder is 12 mm long and 8 mm in diameter, with corner cubes on each end. A small tri-

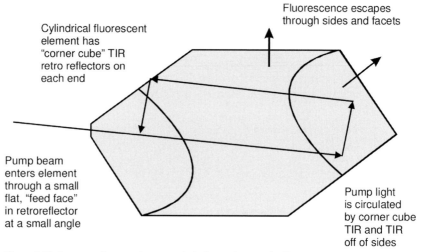

**Figure 5.15** A concept for trapping pump light by total internal reflection.

**Figure 5.16** Rendering of a cooling element with corner cube ends.

angular flat (0.5 mm/side) was assumed to be polished into the tip of the first retroreflector encountered to allow optical pumping at near-normal incidence. The pump beam has a total flux of 1 W, which was distributed amongst 2827 model input rays with 0.8 µm spacing. The pump beam diverges as it travels to the second retroreflector according to its numerical aperture (NA). Each ray is reflected by TIR until it is either absorbed or acquires a position and angle relative to an interface that does not support TIR, in which case it is partially transmitted out of the system.

As the NA was varied, leakage varied as shown in Figure 5.18, with minimum leakage occurring at an NA of approximately 0.05. For very small NAs, a large fraction of the source flux is lost through the entrance face. As the NA increases, fewer rays are lost through the front entrance face, but another leakage mechanism develops that involves "walk off" from the retroreflector facets down to the cylinder wall. The leakage location gradually shifts towards the center of the cylinder with increasing pump NA.

A major leak path is therefore rays walking off a retroreflector facet to the cylindrical portion and leaking out. A single ray is shown in Figure 5.18. Either reducing the entrance flat size by a factor of 2 or doubling the cylinder length significantly decreased pump leakage, as shown in Figure 5.19.

The highest flux concentrations are in the retroreflector components, particularly in the tips. Absorption is much more uniform at the second retroreflector tip than

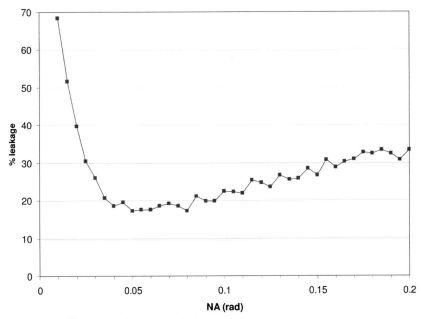

**Figure 5.17** Effect of pump beam NA on leakage.

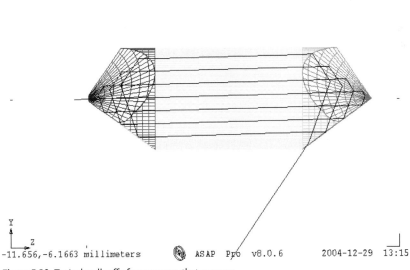

**Figure 5.18** Typical walk-off of a pump ray that escapes.

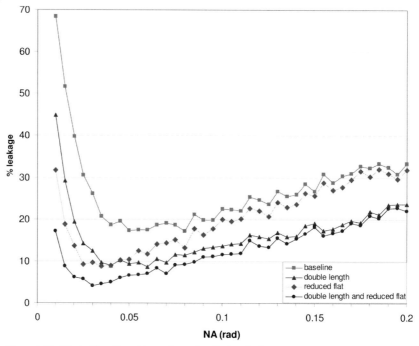

**Figure 5.19** Effect of length and input facet size (flat) on leakage.

the first. This is presumably because flux in the vicinity of the first tip can easily escape, so only the incoming beam with its high flux density results in a significant absorption. The fluence and absorption are also high along the long axis of the device, particularly on the front end. Traces of single rays originating close to the

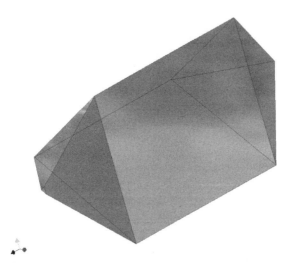

**Figure 5.20** Triangular cylinder cooling element concept.

axis suggest that it is common for a ray to make several cycles through the device and then retrace its path, finally exiting through the entrance face. The high fluence and absorption density on axis may be a result of this.

As a result of the modeling, it was realized that a triangular cylinder might have less pump light leakage than the originally conceived circular cylinder. This would also be easier to interface with the thermal link needed for a refrigerator. However, this concept has yet to be modeled or experimentally verified. A rendering of a triangular cylinder cooling element is shown in Figure 5.20.

An attempt to fabricate retroreflectors from ZBLAN ended in failure when several pieces of ZBLAN broke during fabrication and no more material was then available. The concept is promising but yet to be experimentally verified. Fabrication of a corner retroreflector from Yb:ZBLAN is probably feasible, but will require technicians skilled in working soft, brittle materials and sufficient glass to develop the process.

## 5.6
## Applications

### 5.6.1
### Comparison to Other Refrigeration Technologies

Two different active technologies are commonly used to cool detector focal planes, depending on the temperature required. Thermoelectric coolers (TEC) are capable of cooling focal planes to 180 K starting from an ambient temperature of 300 K. They are small, lightweight, and – being solid state – have no vibration. The main disadvantage of TECs is their efficiency, which drops off rapidly with decreasing temperature, becoming zero around 180 K.

Mechanical coolers, such as Stirling cycle and pulse tube coolers, can produce temperatures below 20 K and are more efficient than TECs. However, mechanical coolers are much larger and heavier than TECs and can produce vibrations that must be canceled or isolated. On spacecraft, focal planes are sometimes cooled with passive radiators, the effective performance of which is highly dependent on mission parameters and the spacecraft's configuration.

### 5.6.2
### Vibration

Because a solid-state diode is used as the pump laser, there are no moving parts and therefore no vibration. This is an obvious advantage for imaging focal planes, particularly when compared with mechanical coolers. TECs also have no source of vibration.

### 5.6.3
### Electromagnetic and Magnetic Noise

The pump laser can be located remotely from the cryocooler dewar, allowing very low levels of electromagnetic interference at the focal plane. This can be important if the focal plane contains EMI or magnetic-field-sensitive devices such as SQUIDs. Optical coolers have this advantage over all other active coolers.

### 5.6.4
### Reliability and Lifetime

An optical cryocooler has no moving parts, which enhances reliability and life. The laser appears to be the lifetime-limiting component. Currently, commercial diode modules have a lifetime of several years in continuous operation. A laser package would consist of many diode laser modules feeding a single output optical fiber using optical Y junctions. This arrangement would have inherent redundancy because the modules almost always fail short and have a distribution of lifetimes that is Gaussian. Additional redundant modules could be added if necessary.

Some mechanical coolers and TECs have proven lifetimes in excess of five years. However, totally redundant coolers can be added only with the use of a heat switch because of the high conductance of these coolers in the off state.

### 5.6.5
### Ruggedness

The cooling element is separated from the heat sink by a gap, and thus is inherently protected from physical stress. The glass, although brittle, has a compact form factor that will allow it to withstand high accelerations. TECs and the cold tips of mechanical coolers must be isolated from loads under acceleration by S-links or other compliant devices.

### 5.6.6
### Cryocooler Mass and Volume

The overall estimated optical cryocooler mass, including the laser, will be smaller than a mechanical cooler for typical focal plane cooling loads. An optical cryocooler, however, will be more massive than a TEC. The lack of vibration allows the cooling element to be closely integrated with the focal plane because *S*-links or other vibration and load isolation devices are not required.

### 5.6.7
### Efficiency and System Mass

We have calculated the power efficiency of an optical cooler based on our photon modeling and information from laser vendors. Solid-state lasers currently have an

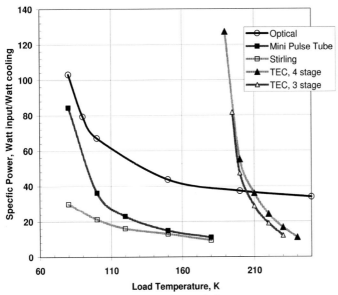

**Figure 5.21** Specific power of cryocoolers, including power supply, with a 300 K sink temperature. The pulse tube cooler data set is for 0.5 to 3.5 W capacity, and the Stirling cooler set is for 2.3 to 7.2 W capacity. The efficiency of mechanical coolers decreases with decreasing size.

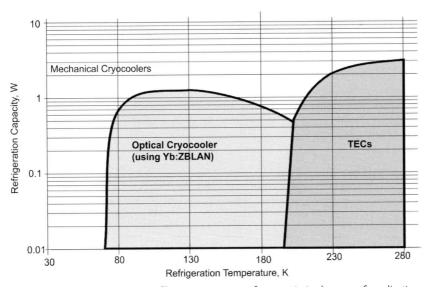

**Figure 5.22** Approximate regions of lowest system mass for an optimized spacecraft application.

efficiency of between 25 and 60% when made up of diode modules. We assumed 50% laser efficiency for the data shown. The calculated specific power of an optical cooler is shown with the measured specific power of other cooling technologies in Figure 5.21.

The lower efficiency of optical cryocoolers and TECs reduces their advantage in power- and mass-sensitive applications such as spacecraft. Mass is required not only for power generation but also to reject the waste heat with a radiator. We used overall system mass data for mechanical coolers using data from Glaister *et al.* [8], but increased the cooling load 20% to account for the cold finger and S-link heat load. For the optical cooler and TEC we used a system penalty of 0.28 kg/W, which is obtained by adjusting data from Glaister and Curran [9] to account for the lower structure and heat transport requirements. Based on cooling temperature and load, regions of minimum system mass were determined and plotted in Figure 5.22. As shown, for spacecraft applications, optical cryocooling will likely have the lowest system mass when the load is less than 1.0 W and the temperature is between 80 and 200 K. Optical cryocooling, in effect, extends the benefits of solid state cooling to this new, lower temperature region.

### 5.6.8
### Cost

Optical cryocoolers will be producible at low cost in volume production; there are no high-precision mechanisms involved, and all of the components used are ones with proven track records in cost-sensitive applications. Mechanical coolers have inherently high costs associated with small tolerance moving parts. High-performance multistage TECs are frequently assembled by hand, but some manufacturers have achieved low costs in volume production of single-stage TECs by using automated assembly. Even the hand-assembled TECs have a large cost advantage over mechanical coolers. Optical coolers are likely to have a similar cost advantage over mechanical coolers.

## 5.7
## Microcooling Applications

C.T.C. Nyugen [10] has suggested developing cryogenic systems for sensors and other small cryogenic devices that significantly reduce the scale of the cold volume and mass, using precisely targeted cooling, MEMS fabrication, and thermal isolation techniques. By reducing the heat leak significantly, the cryogenic payload can be integrated with a miniature cryogenic refrigerator (microcryocooler).

Optical refrigeration appears to be well suited for this approach. The vacuum gap between the cooling element and the heat sink allows the parasitic heat loads due to the refrigerator to be limited to only radiation. At power levels of less than 300 mW, diode lasers are typically less than 1 mm$^3$ in volume (excluding heat sink).

## 5.7 Microcooling Applications | 137

**Figure 5.23** Artist's rendering of a microcryostat containing both the terahertz detector and optical refrigerator beside a quarter. The cryostat has a total volume of 3.0 cm$^3$.

The cooling element volume is inherently small and scales with the required heat lift. The cooling element can be thermally well coupled with the load, minimizing the cold stage mass.

We performed a design study [11] to determine the suitability of an optical refrigerator when using this system concept to produce a highly sensitive portable terahertz detector. A bolometer was modeled with a terahertz antenna and an optical cryocooler into a microcryostat. Figure 5.23 shows the complete system, consisting of the optical refrigerator and bolometer in the microcryostat, with the antenna visible on the outside. The heat leak into the 90 K cold stage of the cryostat is estimated to be 5 mW. The performance was modeled and the results are in the section on modeling. Optical cooling of such a small detector package is feasible provided that cooling material can be produced that has the properties assumed in the model calculations shown in Figure 5.4.

### Acknowledgments

The authors would like to acknowledge the contributions of William Good, John Fleming, Allan Mord, Philip Slaymaker, Jennifer Turner-Valle, Donna Waters and Zongying Wei to this work. The support of the NASA Earth Science Technology Office is also acknowledged.

## References

1 Epstein, R.I., Buchwald, M.I. Edwards, B.C. Gosnell, T.R. and Mugan, C.E. (1995) Observations of Laser-Induced Fluorescent Cooling of a Solid, *Nature*, 377, 500.

2 Edwards, B.C., Anderson, J.E., Epstein, R.I., Mills, G.L. and Mord, A.J. (1999) Demonstration of a Solid-State Optical Cooler: An Approach to Cryogenic Refrigeration, *Journal of Applied Physics*, 86, 6489.

3 Mills, G.L., Fleming, J., Wei, Z. and Turner-Valle, J. (2002) Dielectric Mirror Leakage and its Effects on Optical Cryocooling, *Cryocoolers 12*, Springer, New York.

4 The Essential Macleod v. 8.3, Thin Film Center Inc., Tucson, AZ, 1995–2002.

5 Mansuripur, M. (2001) Omni-Directional Dielectric Mirrors, *Optics & Photonic News*, Sept.

6 ASAP version 7.0.6, Breault Research Organization, Inc., Tucson, AZ, 1982–2001.

7 Mills, G.L., Turner-Valle, J.A. and Buchwald, M.I. (2003) The First Demonstration of an Optical Refrigerator, Cryogenic Engineering Conference, Anchorage, AK, Sept.

8 Glaister, D.S., Donabedian, M., Curran, D.G.T. and Davis, T. (1998) An Overview of the Performance and Maturity of Long Life Cryocoolers for Space Applications, *Aerospace Corporation Report*, TOR–98 (1057)-3, The Aerospace Corp., El Segundo, CA.

9 Glaister, D.S. and Curran, D.G.T. (1996) Spacecraft Cryocooler System Integration Trades and Optimization, *Cryocoolers 9*, Springer, New York, 873–884.

10 Defense Advanced Research Projects Agency/Microsystems Technology (2005) Office Broad Area Announcement (BAA) 05–15, DARPA/MTO, Arlington, VA.

11 Mills, G.L. and Mord, A.J. (2006) Performance Modeling of Optical Refrigerators, *Cryogenics*, 46(2–3), 176.

# 6
# Microscopic Theory of Luminescence and its Application to the Optical Refrigeration of Semiconductors
*Greg Rupper, Nai H. Kwong and Rolf Binder*

## 6.1
## Introduction

Semiconductors are widely thought of as promising candidates for the laser cooling of solid-state systems. Obvious benefits of the use of semiconductors in laser cooling configurations would be their ability to be integrated into existing semiconductor-based optical detection and communications systems. In addition, semiconductors could have fundamental physical advantages. For example, semiconductors are believed to admit cooling down to cryogenic temperatures in the few Kelvins regime. However, even though the basic solid-state laser cooling principles are well understood, and they apply to semiconductors in much the same way as they do to doped glass systems, semiconductors are still posing an enormous challenge. In this chapter, we address the theoretical challenges originating from the fact that semiconductors are complex many-particle systems. In the energy-band picture, the semiconductor consists of many electrons and holes that interact with each other via the Coulomb interaction and that also interact via electron–phonon interactions with the lattice. This results in a conceptual complexity that far surpasses that of atomic and molecular systems.

The basic concept of using luminescence upconversion in order to optically cool solids goes back to [1]. When applied to semiconductors, it can be formulated in the following way. The semiconductor to be cooled is pumped with a narrow-band light beam at the lowest possible transition frequency, which is typically in the vicinity of the lowest bound exciton. In the subsequent thermalization, the optically excited excitons or electron–hole pairs are redistributed among the system's excited states by interacting with phonons. The charge carriers then recombine by luminescing. If the pump (absorption) frequency has been chosen to be below the mean luminescence frequency, the system loses energy in each photon cycle. If not overcompensated for by other (nonradiative for example) loss mechanisms, the system will be cooled.

In contrast to recent successes in the laser cooling of rare-earth doped glasses [2–7] and laser dye solutions [8], experimental challenges to semiconductor laser cooling remain formidable. For this reason, a thorough theoretical understanding

*Optical Refrigeration. Science and Applications of Laser Cooling of Solids.*
Edited by Richard Epstein and Mansoor Sheik-Bahae
Copyright © 2009 WILEY-VCH Verlag GmbH & Co. KGaA, Weinheim
ISBN: 978-3-527-40876-4

of the semiconductor cooling process is especially important. In this chapter, we review our theoretical work, which can be viewed as a logical extension of earlier theoretical investigations. In 1996 and 1997, Oraevsky [9] and Gauck et al. [10] laid the foundations for a rate-equation-based theory in which the effects of nonradiative, radiative and Auger recombination were taken into account. Rivlin and Zadernovsky [11] developed an energy balance theory for semiconductor laser cooling, taking into account the possibility of excitonic luminescence in the limit of vanishing electron–hole (e–h) pair density, but not excitonic absorption. These early theories did not take excitons into account, or they included them in a manner which is only valid for low densities. However, we have found that properly taking excitons into account can lead to significant deviations from simple models, especially for the radiative recombination coefficient. In 2004 Sheik-Bahae and Epstein [12] extended the rate equation approach to include, amongst other things, the effect of luminescence reabsorption. Huang et al. [13, 14] have developed a laser cooling theory that models the differences between the lattice temperature and the plasma temperature, Li [15] has developed a cooling theory for semiconductor quantum wells at the Hartree–Fock level, and Khurgin [16, 17] has proposed to employ surface plasmons on a metal surface close to the semiconductor in order to enhance the cooling efficiency.

Our efforts have focused on developing a microscopic theory of absorption and luminescence and applying it to the feasibility/efficiency analysis of laser cooling of semiconductors. This theory allows us to predict the absorption and luminescence for a large range of e–h densities and a large temperature interval, from a few Kelvins up to room temperature and above. It also enables us to treat doped semiconductors on the same footing. This is important because typical samples used in experiments are unintentionally doped (p-doped, see [18]). Finally, our theory is being generalized to quantum-confined systems, such as semiconductor quantum wells. The microscopic absorption and luminescence results are then used as an input to a generalized version of Sheik-Bahae–Epstein's rate-equation-based cooling theory. Following [12], we calculate the break-even level of the nonradiative recombination as a measure of the cooling threshold requirement.

The major advantage of a good microscopic theory is its predictive power: with a small set of well-established materials parameters, it can provide consistent predictions of many measurable quantities. In doing so, the microscopic theory also yields an understanding of the relevant physics underlying the predicted results. Our absorption/luminescence theory starts with a microscopic Hamiltonian that depends parametrically only on the effective masses of the electrons and the holes and the background dielectric constant, which characterizes the strength of the Coulomb force between the charges. The calculated results, for example the absorption spectra, are traceable consequences of the values of these three parameters and well-specified approximations adopted to make the theory tractable. Thus the theory itself provides the means to estimate the accuracy of the predictions and the directions for systematic improvements. This point is illustrated in the discussions of our calculations below. Overall, we believe that our work represents the most comprehensive microscopic calculations of the luminescence spectra from bulk semiconductors to date.

In past theoretical works on semiconductor laser cooling, there has been significant uncertainty related to the effect of the exciton on the luminescence rate, especially at low temperatures[1]. Our results show that the exciton has highly nontrivial effects on the luminescence spectra and rate: it causes substantial deviation of the density dependence of the luminescence rate from the conventional $BN^2$ (where $B$ is the radiative recombination coefficient and $N$ the electron–hole density) form. Our results also show that, at low temperatures (below 100 K), a theory ignoring the effects of excitons can incur errors that are orders of magnitude in size in the estimates of the cooling threshold (i.e. the break-even nonradiative recombination coefficient).

## 6.2 Microscopic Theory of Absorption and Luminescence

Our microscopic theory of absorption and luminescence is based on a two-band model for the semiconductor, including a valence band and a conduction band. The Coulomb interaction between photoexcited electrons (in the conduction band) and holes (in the valence band) can lead to excitonic resonances. Excitonic resonances play an important role in the cooling process, and therefore it is essential to have them included in the theoretical description. Moreover, since cooling involves optical pumping that creates significant densities of electrons and holes, it is also important for the theory to include a microscopic description of the density-induced modifications of the excitonic resonances, such as density-induced line broadening and frequency shifts. Finally, the theory needs to be able to treat a large range of temperatures, from the few Kelvins regime to above room temperature. In order to fulfill these requirements, we use a microscopic many-particle Hamiltonian within the two-band model as the starting point for our studies. The Hamiltonian consists of several contributions:

$$\hat{H} = \hat{H}_{\text{band}} + \hat{H}_{\text{Coulomb}} + \hat{H}_{\text{electron-light}} . \tag{6.1}$$

Written in terms of creation ($a^\dagger$) and annihilation ($a$) operators for the valence ($\nu = v$) and conduction ($\nu = c$) band electrons, the three terms can be written as

$$\hat{H}_{\text{band}} = \sum_{k\nu} \varepsilon_{k\nu} a^\dagger_{\nu k} a_{\nu k} , \tag{6.2}$$

where $\varepsilon_{k\nu}$ are the parabolic and isotropic energy bands,

$$\hat{H}_{\text{Coulomb}} = \sum_{kk'\nu\nu'} V_q a^\dagger_{\nu k+q} a^\dagger_{\nu' k'-q} a_{\nu' k'} a_{\nu k} , \tag{6.3}$$

---

[1] Excitonic effects in the Auger rate may also be important at low temperatures, but here the Auger rate is highly suppressed. At high temperatures, where Auger recombination is important, exciton effects are expected to be small.

where $V_q = (4\pi e^2)/(v\varepsilon_b q^2)$ is the Coulomb interaction in momentum space ($v$ is the system's volume and $\varepsilon_b$ is its background dielectric constant), and

$$\hat{H}_{\text{electron-light}} = \sum_{k\nu\nu'} d_{\nu'\nu} \cdot E a^\dagger_{\nu'k} a_{\nu k} , \qquad (6.4)$$

where $d_{\nu'\nu}$ is the dipole matrix element (assumed to be $k$-independent and with vanishing diagonal elements; $E$ is the optical field amplitude). We use a classical optical field here, which is appropriate for the evaluation of linear and nonlinear absorption spectra $\alpha(\omega)$. In order to compute luminescence spectra $R(\omega)$, one needs to start out with a quantized light field. However, under rather general assumptions it can be shown that luminescence spectra are related to absorption spectra in the following way, which is called the the Kubo–Martin–Schwinger (KMS) relation (for a proof of this relation, see [19, 20]):

$$R(\omega) = \left(\frac{\omega n_b}{\pi c}\right)^2 \alpha(\omega) g(\omega) , \qquad (6.5)$$

where $g(\omega)$ is the Bose function

$$g(\omega) = \frac{1}{e^{(\hbar\omega-\mu)/(k_B T)} - 1} \qquad (6.6)$$

and $n_b = \sqrt{\varepsilon_b^2}$ is the refractive index (= square root of the background dielectric function), while $\mu$ is the chemical potential (which is strongly affected by interaction effects, including both Coulomb interaction and related exciton effects as well as phonon-induced line broadening). The most important condition for the validity of the KMS relation is that of quasi-thermal equilibrium, which means that the conduction band electrons and the valence band holes are in thermal equilibrium at an arbitrary density and temperature. If the phonon-induced intraband scattering processes are much faster than any of the (radiative or nonradiative) interband excitation and recombination processes, then the quasi-thermal equilibrium regime can be reached with the temperature being that of the lattice[2]. The density in this case is still arbitrary and depends on the optical pumping power and the various recombination processes (see below). The assumption of quasi-thermal equilibrium implies that the charge carriers are distributed according to Fermi functions of their energies (i.e. effects of so-called kinetic holes in the distribution functions are assumed to be negligible). We stress that this does not mean that, in momentum space, the carriers are distributed according to Fermi functions of $k^2$. For example, the conduction-band electron distribution

$$f_e(k) \neq \frac{1}{e^{(\varepsilon_{k,e}-\mu_e)/(k_B T)} + 1} . \qquad (6.7)$$

---

2) In bulk GaAs semiconductors, this is a realistic scenario for a large temperature interval, down to the few Kelvins regime. In GaAs quantum wells, it is believed that quasi-thermal equilibrium can only be reached at sufficiently high temperatures, say above 30 K.

This is because the Coulomb correlations between the charge carriers (i) change the single-particle energy-momentum dispersion relation, and (ii) redistribute the spectral weight of a single-particle momentum eigenstate over a range of energies.

In order to treat these correlation effects – including effects related to exciton formation – in a realistic way, we employ a real-time nonequilibrium Green's function approach. Within this approach, the single-particle, or two-point, fermion Green's functions are defined as follows, where we now use creation and annihilation operators for conduction-band electrons $a_{ck}^\dagger \equiv a_{sk}^\dagger$, $a_{ck} \equiv a_{sk}$ ($s$ labels the two spin states of the conduction band), and valence-band holes $a_{vk}^\dagger \equiv a_{j,-k}$, $a_{vk} \equiv a_{j,-k}^\dagger$ ($j$ labels the quasi-angular momentum states of the valence band):

$$G_n(k, \bar{t}_1, \bar{t}_2) = -i \left\langle T_C \left[ a_{nk}(\bar{t}_1) a_{nk}^\dagger(\bar{t}_2) \right] \right\rangle_0 . \tag{6.8}$$

$T_C$ is the time-ordering operator that orders the operators such that their time arguments increase from right to left, and the contour is a double time contour (the Keldysh contour), which goes from $t = -\infty$ to $t = \infty$ and back to $-\infty$. Explicitly, $T_C[A(\bar{t}_1)B(\bar{t}_2)]$ equals $A(\bar{t}_1)B(\bar{t}_2)$ if $\bar{t}_1$ comes later than $\bar{t}_2$ on the contour $C$, and equals $-B(\bar{t}_2)A(\bar{t}_1)$ if $\bar{t}_1$ comes earlier for any pair of fermion operators $A$ and $B$. $\langle \cdots \rangle_0$ denotes taking the expectation value in the designated quasi-equilibrium state. It is convenient for calculations to write the Green's function in four components according the branch ($C_+$ or $C_-$) in which the time arguments $\bar{t}_1$ and $\bar{t}_2$ reside. Explicitly, write $\bar{t}_1$ as the pair $(t_1, b_1)$ where $t_1$ is the actual time and $b_1 = \pm$ designates the branch (same for $\bar{t}_2$), and the four components $G_n^{b_1 b_2}(k, t_1, t_2)$ are

$$G_n^{++}(k, t_1, t_2) = -i \left\langle T_+ \left[ a_{nk}(t_1) a_{nk}^\dagger(t_2) \right] \right\rangle_0$$

$$G_n^{+-}(k, t_1, t_2) = i \left\langle a_{nk}^\dagger(t_2) a_{nk}(t_1) \right\rangle_0$$

$$G_n^{-+}(k, t_1, t_2) = -i \left\langle a_{nk}(t_1) a_{nk}^\dagger(t_2) \right\rangle_0$$

$$G_n^{--}(k, t_1, t_2) = -i \left\langle T_- \left[ a_{nk}(t_1) a_{nk}^\dagger(t_2) \right] \right\rangle_0 .$$

Here $T_+$ denotes ordinary time ordering and $T_-$ anti-time ordering. We also use the common notations $G_n^<(k, t_1, t_2) \equiv G_n^{+-}(k, t_1, t_2)$, $G_n^>(k, t_1, t_2) \equiv G_n^{-+}(k, t_1, t_2)$, and define the retarded and advanced Green's functions

$$G_n^{R/A}(k, t_1, t_2) = \pm \theta(\pm(t_1 - t_2)) \left[ G_n^>(k, t_1, t_2) - G_n^<(k, t_1, t_2) \right] . \tag{6.9}$$

When the system is in (quasi-)equilibrium, the two-point functions depend only on the relative time $t_1 - t_2$. Under this condition, to obtain all single-particle properties of the system, one only needs to calculate one function for each fermion species – the single-particle spectral function

$$A_n(k, \omega) \equiv -2\mathrm{Im}\, G_n^R(k, \omega) , \tag{6.10}$$

where $\omega$ is the frequency variable conjugate to $t_1 - t_2$. In particular, Re $G_n^R(k, \omega)$ is obtained from the spectral function via the Kramers–Kronig relation, and the other two-point functions are obtained via the corresponding KMS relation [21]:

$$G_n^>(k, \omega) = -i A_n(k, \omega)[1 - f(\omega, T, \mu_n)] \tag{6.11}$$

and

$$G_n^<(k, \omega) = iA_n(k, \omega)f(\omega, T, \mu_n) ,\qquad(6.12)$$

where

$$f(\omega, T, \mu_n) = \frac{1}{e^{(\hbar\omega-\mu_n)/k_B T} + 1} ,\qquad(6.13)$$

which is the fermion distribution function at temperature $T$ and chemical potential $\mu_n$.

For a given approximation scheme, a diagrammatic representation of the equation of motion of the Green's function can be realized (see for example [22–24]). In the following we use a $T$ matrix approximation, which means that the self-energy entering the equation for $G$ (the Dyson equation) involves an infinite sum over ladder-type diagram. In Figure 6.1a,b the Dyson equation and the corresponding $T$-matrix equation is shown. The Green's functions can be either electron ($G_e$) or hole ($G_h$) Green's functions.

For the single-particle electron and hole self-energies, the theory contains ladder-type (screened $T$-matrix) diagrams involving the same species ($T_{ee}$ and $T_{hh}$) as well as different species ($T_{eh}$ and $T_{he}$) (second term on the right-hand side in Figure 6.1a). We also include the exchange self-energy in the first order of the Coulomb potential, the third term on the right-hand side in Figure 6.1a, which is called Hartree–Fock (HF) self-energy and is formally the exchange diagram corresponding to the $T$ self-energy (neglecting second- and higher-order contributions), as well as the corresponding and Coulomb-hole (CH) [25] contributions.

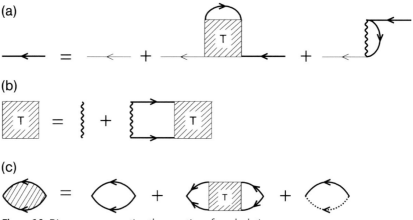

**Figure 6.1** Diagrams representing the equations for calculating the single-particle Green's functions and the optical susceptibility. (a) Thick line: full single-particle Green's function. Thin line: noninteracting Green's function. Wavy line: screened Coulomb potential. Hatched box: $T$ matrix. (b) Ladder diagrams for the $T$ matrix. (c) Ladder approximation for the susceptibility (hatched bubble). Dashed line: acceptor Green's function. From [65].

In Figure 6.1c we show the optical susceptibility $\chi(\omega)$ from which the absorption coefficient $\alpha(\omega)$ follows:

$$\alpha(\omega) = \frac{4\pi\omega}{n_b c} \operatorname{Im}\left[\chi^R(q = \omega n_b/c, \omega)\right]. \tag{6.14}$$

The last contribution in Figure 6.1c describes optical transitions between the conduction band and impurity states. We will neglect this term for now, and come back to it at the end of this chapter (see Section 6.6).

The inclusion of the $T$ self-energy, which is nonperturbative (in the Coulomb interaction), has the important consequence that exciton states (electron–hole pairs bound by the Coulomb interaction) are included in the quasi-particle renormalization. Hence, we are able to describe a system of partially ionized excitons; in other words, one in which unbound electrons and holes (an "electron–hole plasma") coexist with excitons. This theoretical description of a system in which the e–h pairs form a partially ionized exciton gas is similar to the approach used for bulk semiconductors in [26] and for quasi-one-dimensional semiconductors in [27,28]. In addition to the renormalization of the single-particle states via the $T$ self-energy, the electron–hole $T$-matrix $T_{\mathrm{eh}}$ determines the optical interband susceptibility [19, 29, 30] (see Figure 6.1c). This allows for the description of excitons as well as unbound (e–h continuum) states in the optical spectra. The e–h $T$ matrix is a ladder-type sum of e–h Coulomb interaction lines, and we use quasi-statically screened e–h Coulomb interaction lines [25] in the calculation of all $T$ matrices (both electron–hole as well as electron–electron and hole–hole). Whereas an unscreened $T$-matrix theory is applicable only in the low-density regime, the inclusion of screening in the $T$ matrix also renders the theory valid in the high-density limit, in which plasma screening changes the nature of the Coulomb interaction from long range to short range. In this high-density limit, no bound states exist, and correspondingly the screened $T$ matrix theory does not exhibit any exciton resonances (see also [31]).

From the normalization of the Green's functions to the given electron (hole) densities $N_e$ ($N_h$), we find the chemical potentials $\mu_e$ ($\mu_h$), which are the chemical potentials for the interacting electrons (holes). The chemical potentials require a self-consistent evaluation, because they depend on the single-particle Green's functions and the $T$ matrices, which in turn depend on the chemical potentials.

The e and h spectral functions are given by the imaginary parts of the retarded Green's functions. They enter the calculations of the chemical potentials, and are different from those for noninteracting particles (i.e. they are not just delta functions in frequency space peaked at the single particle energies). In our case, they contain features due to excitonic as well as plasma-induced renormalizations. The interaction-induced features (more or less well separated peaks) have been associated with populations of carriers bound in excitons [19]. This yields, at least conceptually, a way to distinguish free (or unbound) from bound charge carriers, which is important, because free and bound carriers contribute differently to the screening of the Coulomb potential. The Coulomb potential entering all self-energies and $T$ matrices is screened mainly by the unbound carriers (plasma). In the following, we specify our approach to the estimate of the plasma fraction or ionization ratio (the ratio of plasma density over total density), which is used to determine the screening within the quasi-static plasmon-pole approximation [25].

Since, in practice, the spectral function cannot always be unambiguously decomposed into contributions from bound and unbound carriers (see [32] for details), we use a simplified approach. We define the ionization degree as $N_{\text{free}}/N$, where $N_{\text{free}}$ is approximated by the density of a hypothetical noninteracting plasma at the same chemical potential as that obtained for the actual density $N$ of the interacting plasma. The single-particle energies in the hypothetical plasma include only HF and CH shifts. Numerical results for the ionization ratio are shown in Figure 6.2 as a function of total density for various temperatures. For all temperatures shown, in the low-density limit, our numerical results are in agreement with the mass-action law given by the well-known Saha equation (according to which the exciton density is proportional to the product of the e and h densities) and its generalization, the Beth–Uhlenbeck formula [33] (see also [34–36]). In the high-density limit, our theory yields complete exciton ionization, which is often referred to as the Mott transition. At low temperatures, for which $k_B T$ is less than the exciton binding energy (in our case, temperatures of less than approximately 50 K), we find an intermediate range of densities where the ionization ratio is small, indicating the existence of excitons.

We see from Figure 6.1a that the $T$ matrix self-energy is a convolution of the $T$ matrix with the single-particle Green's functions. We can write this schematically as $\Sigma = TG$. Physically, this means that a carrier propagating through the system, whose properties are renormalized by $\Sigma$, is interacting via the $T$ matrix with other carriers (represented by $G$), which in turn are also interacting with other carriers through the renormalization of $G$. From a numerical feasibility point of view it is very challenging to include these self-energies. It is therefore often useful to introduce a simplified self-energy, for example one in which the Green's function multiplying $T$ is replaced by a noninteracting Green's function $G^{(0)}$. In the frequency domain, the imaginary part of $G^{(0)}$ is proportional to a delta function peaked at the quasi-particle energies (i.e. the energies of the particle renormalized only by the static Hartree–Fock and Couloumb hole self-energies). In the next section, we will compare some of our cooling criteria obtained from the full $T$ matrix calculations using $G$ and $G^{(0)}$, respectively. We will denote those two approximations as "$TG$"

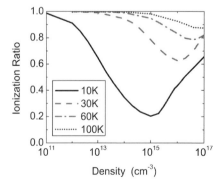

**Figure 6.2** Ionization ratio for various temperatures as a function of density. From [32].

and "$TG^{(0)}$" approximations, respectively. While the "$TG$" and "$TG^{(0)}$" approximations generally yield different results as far as absorption and luminescence spectra are concerned (this is especially true at low temperatures), the cooling results presented below will show far fewer discrepancies between those two approximations. The underlying reason for this is that the cooling results involve spectrally integrated quantities. Since the "$TG$" and "$TG^{(0)}$" are found to yield very similar results as far as the cooling analysis is concerned, we will provide most of the cooling data in the "$TG^{(0)}$" approximation. We also note that the speed of the calculation can be enhanced with another approximation. The free two-particle Green's function $G^{(2)}(k, q, \omega)$, which enters the calculation of the $T$ matrix and the susceptibility, depends on both the center of mass momentum $q$ and the relative momentum $k$. The overall calculation can be simplified by accounting for the center of mass momentum as a simple energy shift $g(k, q, \omega) = g(k, q = 0, \omega - (\hbar q^2)/(2m_r))$. This is strictly valid only in the absence of Pauli blocking and broadening of the spectral functions. We use this approximation in the $TG^{(0)}$ model. In the $TG$ model we calculate the true $q$ dependence, but with the additional simplification of angle averaging, which removes the angle dependence between $k$ and $q$. More details will be published in [32].

As pointed out above, the description of a partially ionized exciton gas requires the inclusion of the $T$ self-energy (or any generalization thereof); an approximation at the Hartree–Fock is not sufficient in this case. However, certain aspects of the $T$ matrix approximation must be viewed with caution. When calculating the ionization degree it is appropriate to use the full $T$ self-energies, because the ionization degree follows from the single-particle spectral function. On the other hand, we note that there are indications that the corresponding (excitonic) vertex corrections may be important for the optical spectra [19, 37]. Unfortunately, such vertex corrections are presently not numerically feasible. Since it is reasonable to expect that contributions from excitonic resonances to the e–h $T$ matrix self-energy are partially canceled out by the vertex corrections, we proceed in the following pragmatic way. We calculate the cooling characteristics twice: first, using the full theory, which includes the $T$ self-energies as discussed above; second, using a theory in which, after the self-consistent solutions for $T$ self-energies and the Green's function are obtained, the solution cycle is iterated one more time for the Green's function, but this time with the $T$ self-energy evaluated only to second order in the screened Coulomb interaction. This Green's function is then used to obtain the susceptibility. The second-order $T$ self-energy does not contain excitonic resonance. In other words, within the "second-order" model we use the full $T$ matrix self-energies basically only for the calculation of the chemical potential and the ionization degree; the ladder diagrams for the susceptibility contain self-energy insertions only up to second order in the Coulomb interaction. We then compare our most essential cooling results obtained using the "full" theory with those obtained with the "second-order" model. We believe that the true cooling results are between these two sets of data obtained within the "full" and "second-order" theories. As in the "full" theory, we also specify in the "second-order" model whether we use $G$ or $G^{(0)}$ in the calculation of the self-energy. Below, we will use the labels $TG$, $TG^{(0)}$, $T^{(2)}G$ and $T^{(2)}G^{(0)}$

to denote the "full" $T$ matrix theory using the full $G$, the "full" $T$ matrix theory using $G^{(0)}$, the "second-order" model using the full $G$, and the "second-order" model using $G^{(0)}$, respectively.

The $T$ matrix theory outlined above accounts for the modification of optical spectra through e–h pairs. In the limit of vanishing e–h density, the spectra are dominated by the coupling of electrons and holes (or of excitons) to phonons as well as impurities. These effects determine, amongst other things, the excitonic lineshapes in absorption and luminescence. The temperature-dependent (zero-density) bandgap renormalization is taken into account using literature values [25] for the temperature-dependent bandgap energy

$$E_g = 1520 \text{ meV} - 0.56 \frac{T^2}{T + 226 \text{ K}} \text{ meV/K} .$$

When using luminescence spectra in a cooling analysis, it is very important to account for the non-Lorentzian exciton lineshape, which, in the zero-density limit, results mainly from interactions with phonons. As for the temperature-dependent zero-density exciton lineshape, we use a phenomenological lineshape consistent with experimental data [38]. The temperature-dependent linewidth has contributions from the interaction of electrons with longitudinal acoustic (LA) and longitudinal optical (LO) phonons. Its temperature dependence can be modeled (see [38]) as

$$\gamma_{ph}^0 = \gamma_{LA} T + \frac{\gamma_{LO}}{e^{\frac{\Omega_{LO}}{k_B T}} - 1} , \quad (6.15)$$

where, for GaAs, we use $\gamma_{LA} = 3.9 \, \mu\text{eV}$, $\gamma_{LO} = 30.4$ meV, and $\Omega_{LO} = 36$ meV[3]. If the phonon-induced broadening is assumed to be frequency-independent, one obtains a Lorentzian lineshape for the absorption spectrum. In such a model, the absorption decreases as $\sim 1/\omega$ at frequencies much less than the 1 s exciton frequency. An artefact of that model stems from the fact that, when used with the KMS relation to calculate the luminescence spectra, the Lorentzian tail does not drop off fast enough. The multiplication with the exponential in the Bose function results in a spurious increase in luminescence at low energies, which is particularly damaging in a theory aimed at the understanding of cooling, since the spurious low-frequency tail distorts any estimate of the mean luminescence frequency. It is well known that the low-frequency drop-off of a phonon-broadened lineshape should be exponential asymptotically. At elevated temperatures, this exponential tail is called the Urbach tail (see for example [39, 40]). At low temperatures, the exponential tail can exhibit discrete features due to LO-phonon sidebands, but in GaAs the sidebands do not overwhelm the general trend of an approximate exponential decrease in absorption far below the exciton resonance [41].

---

3) Gopal [38] used a parameter of $\gamma_{LA} = 13 \, \mu\text{eV}$. Gopal's data, used a 3 μm thick sample, but did not appear to account for propagation effects. After accounting for propagation effects, we believe that $\gamma_{LA} = 3.9 \, \mu\text{eV}$ provides a better fit to the data.

In order to obtain the desired exponential decay in the phonon-broadened absorption spectrum, one needs to introduce a frequency dependence into the dephasing width. In doing this, we impose three requirements which admit the desired properties and at the same time preserve certain physical consistency conditions.

(i) In order to properly model the dephasing of the exciton peak, the phonon broadening is added to the two-particle (electron–hole) Green's function, rather than the single-particle Green's function.

(ii) To ensure causality, the frequency-dependent phonon width is added as the imaginary part of a self-energy to the electron–hole Green's function. The corresponding real part is obtained by the Kramers–Kronig relation. The additive constant in the Kramers–Kronig relation is chosen such that the shift of the lowest exciton due to the phenomenological phonon-induced self-energy vanishes.

(iii) With the phonon-induced self-energy included, the resulting absorption spectrum must still cross zero at the chemical potential so as to preserve the positivity of the luminescence spectrum.

We construct the phonon-induced width as follows. First, to get an exponential tail at frequencies far below the exciton resonance, we define

$$\gamma'_{ph}(\omega) = \frac{2\gamma^0_{ph}}{1 + e^{a(\omega-\varepsilon_0)/(k_B T)}} . \tag{6.16}$$

Here, $\varepsilon_0$ is set to be the lowest exciton energy ($\varepsilon_0 = E_g - E_R$) at all densities that are low enough so that the bandgap shift (which we define to be Coulomb hole self-energy plus the sum of the electron and hole exchange self-energies evaluated at zero electron and hole wavevectors) has not reached the lowest exciton energy. At high densities, where the bandgap has shifted below the exciton energy, we set $\varepsilon_0$ at the bandgap energy. The parameter $a$ determines the slope of the low-energy tail, and was adjusted to fit the data shown in [38]. We found that $a = 4$ provides a reasonable fit for all temperatures. To ensure that the absorption crosses zero at the chemical potential, it is sufficient to make the phonon width cross zero at the same point. This can be achieved by defining

$$\gamma_{ph}(\omega) = \gamma'_{ph}(\omega)\left(1 - \frac{2}{e^{(\hbar\omega-\mu)/(k_B T)} + 1}\right) . \tag{6.17}$$

This expression for the width is used as the imaginary part of a phonon-induced self-energy of the electron–hole Green's function. The corresponding absorption and luminescence spectra in the low-density limit are shown in Figure 6.3 for several different temperatures. At a sufficiently low temperature (30 K) we can clearly see the 1 s exciton peak and the e–h continuum. At higher temperatures, the sharp exciton resonance vanishes. In Figure 6.4 we show low-temperature luminescence spectra for various densities. As expected, the sharp exciton resonance vanishes at higher densities (Mott transition).

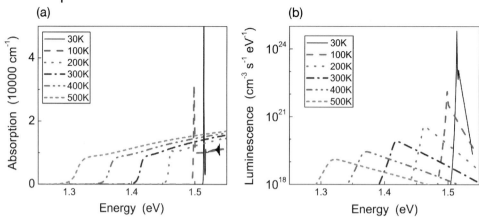

**Figure 6.3** (a) The calculated absorption spectra for $T = 30$, 100, 200, 300, 400 and 500 K at a density of $10^{14}$ cm$^{-3}$. (b) The corresponding luminescence spectra. From [46].

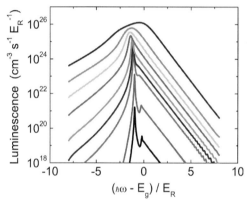

**Figure 6.4** Luminescence spectra at $T = 30$ K for various densities: *from bottom to top* $10^{13}$, $10^{14}$, $10^{15}$, $2 \times 10^{15}$, $4 \times 10^{15}$, $10^{16}$, $2 \times 10^{16}$, $4 \times 10^{16}$, $10^{17}$ cm$^{-3}$. From [32].

We finally note that the theory formulated above does not take into account light propagation effects, such as luminescence reabsorption, polariton effects and partial or total internal reflection at the semiconductor-to-air interface. We are presently investigating those effects on the basis of a theory that involves the photon Green's function in addition to the electronic Green's functions used above. Initial results indicate that, for the range of temperatures and densities that we are interested in, luminescence reabsorption and interface reflection lead to a slight spectral reshaping of the luminescence spectra, with a significant overall loss of luminescence intensity. In the cooling theory discussed below, such an overall loss is modeled by a phenomenological factor, the luminescence extraction efficiency $\eta_e$. Our generalized theory involving the photon Green's function is able to predict $\eta_e$, but in the following we restrict ourselves to the case where it is regarded as a phenomenological parameter. Initial results including polariton effects indicate that the luminescence spectra inside the semiconductor can vary significantly at very low temperatures (< 20 K) and very low densities. In the density regime relevant for cooling, which is below but close to the Mott density, polariton effects seem to be negligible. Interestingly, even in the low-temperature low-density case, where polariton effects in the vicinity of the 1 s exciton change the luminescence spectra inside the sample, it appears that the spectra outside the sample are still similar to those calculated without polariton effects. This is in qualitative agreement with similar findings in [42]. More details on the luminescence theory including light propagation effects will be published in [32].

## 6.3 Cooling Theory

In this section, we present our approach to the theory of the optical refrigeration of semiconductors. Our theory is based on – but represents a significant extension of – the rate-equation approach of [12]. There, it was shown that cooling can only occur if the energy removal from the system (which happens in the absorption–luminescence cycle) dominates over all other loss (heating) mechanisms, including nonradiative recombination, luminescence reabsorption, Auger recombination, free-carrier absorption and other parasitic absorption processes. The net power imparted to the system is given by

$$P_{net} = I[\alpha(\omega_a, N) + \alpha_b + \sigma_{fca} N] - \hbar\omega_l \eta_e BN^2 , \qquad (6.18)$$

where, at steady state, the e–h pair density $N$ is given by

$$\frac{dN}{dt} = \frac{\alpha(\omega_a, N)}{\hbar\omega_a} I - AN - \eta_e BN^2 - CN^3 = 0 . \qquad (6.19)$$

Here, $I$ is the intensity of the pump laser, $\alpha_b$ is the (parasitic) background absorption, $\sigma_{fca} N$ is the free carrier absorption, $AN$ is the nonradiative recombination rate, $\eta_e$ is the extraction efficiency (the probability that an emitted photon escapes from the semiconductor without being reabsorbed), $BN^2$ is the radiative recombination rate, and $CN^3$ is the Auger recombination rate.

The simple density dependence (linear for nonradiative, quadratic for radiative, and cubic for Auger recombination) is not always correct. We will see below that the simple model in which the radiative recombination $\sim N^2$ is only valid for low densities. Many-particle effects, such as excitonic effects, can lead to considerable deviations from the $N^2$-dependence. Even phase-space filling effects, which are of course included in our theory, can do this [43]. To correctly model the radiative recombination for all densities, we introduce a density-dependent radiative recombination coefficient,

$$B(N, T)N^2 = \int \frac{d\omega}{2\pi} R(\omega, N, T) \equiv L(N, T) , \qquad (6.20)$$

where the luminescence $L$ and correspondingly the density-dependent radiative recombination coefficient $B$ is calculated using our microscopic theory.

The Auger recombination rate is also likely to be different from the simple $CN^3$ model for very low temperatures and/or very high densities. For example, it has been shown that deviations at high densities can be caused in two-dimensional systems by Coulomb screening and phase-space filling [44]. Also, in one-dimensional systems, it has been shown that the Auger recombination rate for excitons is proportional to the square of the exciton density [45]. However, possible similar deviations in the present case of bulk GaAs are not likely to have a significant effect on the cooling results in this paper. We have found that Auger recombination is only important to the cooling results for temperatures above 200 K [46]. In order for cooling to be possible, the Auger recombination rate must be significantly smaller then the radiative recombination rate. This is only possible if the density is kept reasonably small (compare Figure 6.8 below). With these two restrictions, the $CN^3$ model for Auger is expected to produce reasonable results.

Cooling requires $P_{net} < 0$. In an experiment, a desired starting temperature ($T$) is chosen. Then $A$, $C$, $\eta_e$, $\alpha_b$ and $\sigma_{fca}$ are fixed parameters, while the intensity (and thereby $N$) and $\omega_a$ are variables that can easily be adjusted. A useful cooling criterion defined in [12] is the break-even nonradiative recombination coefficient, $A_b$, which marks the boundary at which cooling becomes possible ($P_{net}(A_b) = 0$). The break-even $A_b$ is given by

$$A_b = \frac{\hbar\omega_l \eta_e B(N) N}{\hbar\omega_a \left(1 + \frac{\alpha_b + \sigma_{fca} N}{\alpha(\omega_a, N)}\right)} - \eta_e BN - CN^2 \; . \tag{6.21}$$

A numerical maximization is then used to find the optimal $A_b$ over the space of all densities and absorption frequencies,

$$\bar{A}_b = \max A_b(N, \omega_a) \; . \tag{6.22}$$

We denote the density and absorption frequency necessary to achieve the maximum $A_b$ (denoted by $\bar{A}_b$) by $\bar{N}$ and $\bar{\omega}_a$, respectively. Physically, $\bar{A}_b$ represents a requirement on the sample quality needed to achieve cooling in an experiment with the fixed parameter values listed above.

In the following discussion, we use (unless specified otherwise) the parameters $\sigma_{fca} = 10^{-20}$ cm$^2$, $\alpha_b = 1$ cm$^{-1}$[4], $\eta_e = 0.25$, and a temperature-dependent Auger coefficient [12, 18]

$$C(T) = C(300\text{ K})\exp(2.24(1 - 300/T[\text{K}])) \; ,$$

with $C(300\text{ K}) = 4 \times 10^{-30}$ cm$^6$s$^{-1}$ and the $T$ values specified below.

---

4) The background absorption is assumed to be dominated by the absorption of the dome (e.g. ZnSe), which is assumed to have $\alpha L = 10^{-4}$. Modeling this absorption as a background absorption inside the active medium (GaAs) and assuming it to be 1 μm thick, we obtain $\alpha_b = 1$ cm$^{-1}$.

## 6.4
## Cooling of Bulk GaAs

The Sheik-Bahae–Epstein (SB–E) theory [12] models the luminescence rate as $BN^2$, where $B$ is a constant. As pointed out above, see (6.20), in a general theory the radiative recombination coefficient $B$ is density dependent. We study the density dependence of $B$ in Figure 6.5 using various models. One group of models contains exciton effects (labeled "E"), while the other group (labeled "F") does not contain any e–h Coulomb interaction and hence contains no exciton effects. In the first group, we have the full $T$ matrix calculation evaluated with $G^{(0)}$; this model contains the exciton resonance in the spectra as well as exciton occupation effects, described by the (excitonic) correlation peak in the renormalized e and h spectral functions, as well as plasma screening due to the plasma fraction of the partially ionized e–h system. The corresponding second-order approximation does not contain the correlation peak in the spectral functions. Finally, this group includes what is sometimes referred to as "plasma theory" [25, 30, 47–49], which contains exciton resonances in the optical spectra, but the renormalizations of the particle properties at nonzero e–h densities are restricted to frequency-independent plasma contributions (technically, screened Hartree–Fock self-energies), in our case the quasi-statically screened Hartree–Fock and Coulomb-hole self-energies. It also contains the effects of quasi-static screening of the Coulomb interaction, but unlike the "full" and "second-order" $T$ matrix calculations, the screening here is caused by the entire e–h system. The "plasma theory" (which we label "SHF" for screened Hartree Fock) does not account for plasma-assisted dephasing of the exciton resonance. Rather, the width of the exciton resonance is determined only by the density-independent phonon-assisted dephasing. All models within this group also include phase-space filling. In the second group, labeled "F", no Coulomb interaction between electrons and holes is included. The "free" model includes only phase-space filling; it simulates a system where the carriers do not interact with each other or with phonons, and where the Pauli exclusion principle is the only cause of density-dependent optical properties. The second group also contains a model where the carriers are assumed not to interact with each other, but where phonon-induced dephasing is taken into account (labeled "free/ph").

The most obvious conclusion to be drawn from Figure 6.5 is the fact that excitonic effects enhance the radiative recombination coefficient. This is especially true at low temperatures. Figure 6.5a shows an enhancement of almost two orders of magnitude at 30 K. Further studies have shown that $B$ increases almost exponentially with decreasing temperature for fixed and sufficiently small density (see [46]).

From general physical arguments, we expect the following behavior for the density dependence of $B$. For low densities, the system is composed of a dilute e–h gas. In this regime, the electron must first find a hole before it can recombine and emit a photon. Hence, the luminescence rate is proportional to $N^2$ and $B(N)$ is a constant. If the system is composed of an exciton gas, the electron is bound to the hole and the luminescence rate is proportional to $N$. If the system is sufficiently dense to be a degenerate two-component Fermi fluid, then band filling

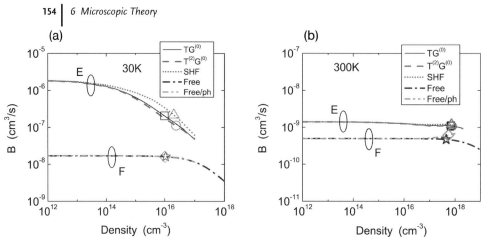

**Figure 6.5** The density-dependent radiative recombination coefficients for (a) 30 K and (b) 300 K calculated using several different models. Group "E" comprises several excitonic models: full $T$ matrix ($TG^{(0)}$), second-order $T$ matrix ($T^{(2)}G^{(0)}$), and screened Hartree–Fock (SHF) model. Group "F" contains models without e–h Coulomb interactions: "free" and "free with phonons" ("Free/ph"). From [46].

again leads to a luminescence rate proportional to $N$ [43]. Figure 6.5 shows that, as the density becomes small, $B(N)$ approaches a constant for all models as expected. As the temperature increases, the transition density between the dilute gas regime and the dense gas regime also increases. At 300 K, the system has not transitioned into a dense gas by $10^{18}$ cm$^{-3}$, and $B(N)$ is a constant for all densities shown. At 30 K, we see that the excitonic models all have significantly higher B than the free models. As, with increasing density, excitons start to form at densities $\sim 10^{14}$ cm$^{-3}$, B begins to drop off. The full and second-order theories produce nearly identical results. For any given density, B is higher for the plasma (SHF) theory. This is due to the neglect of the excitonic correlation contributions to the single-particle spectrum, which increases the chemical potential $\mu$ and consequently B. The symbols in Figure 6.5 mark the optimal cooling density, which will be discussed below. For low temperatures, the B value at the optimal cooling density is significantly less then the low-density value.

We note that, even though the ionization degree approaches unity as $N \downarrow 0$ (in agreement with the Saha equation), the excitonic and plasma contributions to the luminescence remain of the same order in the free density in this limit. The exciton luminescence is proportional to the exciton density, and can be written as $L = B_{\text{exc}} N_{\text{exc}}$, while the plasma luminescence is proportional to the square of the free carrier density, $L = B_{\text{free}} N_{\text{free}}^2$. However, from the Saha equation, $N_{\text{exc}} \propto N_{\text{free}}^2$, while in the zero-density limit $N_{\text{free}} \sim N$. Hence, both the exciton and plasma have an approximate $N^2$ dependence in the zero-density limit.

We turn now to the cooling analysis. As mentioned above, the optimal density $\bar{N}$ and pump frequency $\bar{\omega}_a$ follow from the optimization of (6.21). The optimal pump frequency, together with the mean luminescence frequency,

$$\omega_\ell = \frac{\int d\omega\, \omega R(\omega)}{\int d\omega R(\omega)},$$

yield the upconversion $\Delta\omega = \omega_\ell - \tilde{\omega}_a$. In order to enhance our understanding of $\Delta\omega$, for which the density is fixed at the optimal density $\tilde{N}$, we discuss in the following a density-dependent function $\Delta\tilde{\omega}(N)$, which is designed to coincide with $\Delta\omega$ for $N = \tilde{N}$. It is defined as $\Delta\tilde{\omega} = \omega_\ell - \tilde{\omega}_a(N)$ where $\tilde{\omega}_a(N)$ maximizes the break-even nonradiative decay rate $A_b$ over the space of all absorption frequencies $\omega_a$ for fixed $N$; in other words, it solves

$$\bar{A}_b(N) = \max A_b(\omega_a)|_N . \tag{6.23}$$

It can be seen from Figure 6.6 that $\Delta\tilde{\omega}$ has a peak at moderate densities. At low densities, $\omega_\ell$ is slightly above the exciton resonance, and – generally – the stronger the exciton resonance the closer $\omega_\ell$ is to the resonance. As the density increases, the exciton strength decreases, causing $\omega_\ell$ and therefore $\Delta\tilde{\omega}$ to increase. At high densities, $\Delta\tilde{\omega}$ quickly decreases and becomes negative. This is due to the requirement $\alpha(\omega_a) > 0$. As discussed earlier, $\mu$ marks the boundary between absorption and gain, so $\omega_a > \mu$. $\mu$ increases with increasing density. It can be shown that the sudden reduction in $\Delta\tilde{\omega}$ occurs when $\mu$ gets close to $\tilde{\omega}_a$ [46]. At this point, $\tilde{\omega}_a$ must increase in order to stay in the absorption region, and cooling becomes impossible when $\tilde{\omega}_a > \omega_\ell$.

Figures 6.7 and 6.8 show the optimal $\bar{B}$ and $\tilde{N}$, which enter into the break-even cooling calculation for the optimized break-even nonradiative lifetime $\bar{\tau}_b$. The latter is defined as $\bar{\tau}_b = 1/\bar{A}_b$ and is shown in Figure 6.9. The solid and dotted lines show the results of the full calculation and the second-order approximation using $G^{(0)}$ in the $T$ self-energies. The overall temperature dependence of $\bar{\tau}_b$ is similar for these

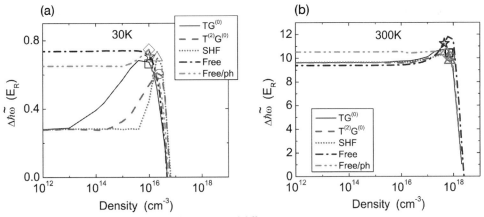

**Figure 6.6** Upconversion $\Delta\hbar\tilde{\omega}$, calculated using several different models, for 30 K (a) and 300 K (b). The optimal cooling density $\tilde{N}$ has also been marked. From [46].

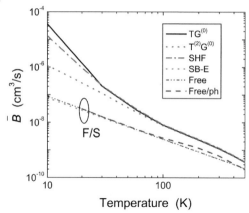

**Figure 6.7** The radiative recombination coefficient ($B$) at the optimal cooling density for several different theories. The group labeled "F/S" contains the "free", "free with phonons" and SB–E models. From [46].

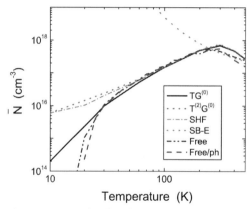

**Figure 6.8** Optimal cooling density vs. temperature for several different theories. From [46].

two curves[5], suggesting that many-body correlation effects beyond the $T$ matrix approximation would probably not change the results significantly. Comparing the "$TG^{(0)}$" with the "$TG$" model, we see that the inclusion of correlations in the single-particle Green's function that enters the $T$ matrix self-energies does not affect the results significantly. Note that, since the numerical challenge increases with decreasing temperature, we only show the "$TG$" results for temperatures above 30 K. The absence of data below 30 K does not imply that cooling is impossible in this region.

---

5) The luminescence spectra in the full $T$ matrix and second-order approximations are quite different, but the resulting differences in the values of the optimal $N$ and $B$ largely compensate each other, leading to similar values for $\bar{\tau}_b$.

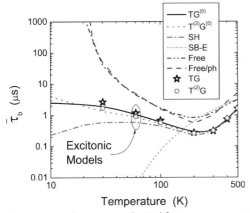

**Figure 6.9** Break-even nonradiative life time vs. temperature for several different theories. For each model, the predicted cooling region is above its curve. From [46].

We also show $\bar{\tau}_b$ for several simplified models. In addition to the simplified models already discussed above, we present results from a model that is based on an analytical expression derived by Sheik-Bahae and Epstein in [12], which we label as "SB–E" model. In order to evaluate the SB–E model in a way suitable for our calculations, we use slightly different parameter values from [12], namely $B(300\,\text{K}) = 4.69 \times 10^{-10}\,\text{s}^{-1}\text{cm}^{-3}$, $\hbar\Delta\omega = 1.55\,k_B T$ (note that in the SB–E model the upconversion is a parameter rather than an optimizable function). In this way, the SB–E model coincides with our "free" model if we neglect phase-space filling in the "free" model. In [12], background absorption is included in the general formalism but not in the temperature-dependence analysis. The SB–E model used here does not include background absorption. We also note that [12] already contains a qualitative discussion of the influence of phase-space filling, which we find to be in agreement with our quantitative results obtained using the "free" and "free with phonons" models.

In the following, we will analyze the basic features of $\bar{\tau}_b$ and try to further elucidate the physics governing the temperature dependence of $\bar{\tau}_b$. In the basic SB–E theory,

$$P_{\text{net}} = AN\hbar\omega_a - \eta_e BN^2 \Delta\hbar\omega + CN^3 \hbar\omega_a \,. \tag{6.24}$$

One sees that increasing the density allows the radiative recombination (cooling) to dominate over the nonradiative recombination (heating). Auger recombination (heating) limits the maximum density. As the temperature decreases, the Auger recombination coefficient decreases exponentially. In the SB–E theory, this allows the optimal cooling density to increase, causing $\bar{\tau}_b$ to decrease exponentially (see Figures 6.8 and 6.9). In more realistic models, band-filling effects limit the optimal cooling density. As the density increases, the chemical potential eventually forces $\omega_a > \omega_\ell$, thus preventing cooling. The free e–h theory adds band-filling effects to the SB–E theory. In Figure 6.9 we can see that the free e–h theory reproduces the results from the SB–E theory for temperatures above 200 K. Below 200 K, the free e–h theory results in dramatically larger $\bar{\tau}_b$. Figure 6.8 shows that the increase in $\bar{\tau}_b$ below 200 K is due to band-filling effects which limit $\bar{N}$ and cause $\bar{N}$ to decrease with decreasing temperature.

Figure 6.9 shows that excitonic effects make it easier to achieve cooling for any initial starting temperature. For approximately $T \geq 40$ K, the improvement comes from the increased radiative recombination rate $B$. Figure 6.7 shows that the exciton resonance increases $B$ over the full temperature range shown. This increase in $B$ takes place even though, at the optimal cooling density, $B$ is generally less then its low-density value.

At very low temperatures, $T \leq 40$ K, the background parasitic absorption plays a significant role in cooling. In fact, for the free e–h theory, the background absorption prevents cooling for $T \leq 20$ K. This is indicated by a negative $\bar{A}_b$ and correspondingly the absence of a physical solution for $\bar{\tau}_b$. A detailed analysis of the interplay between the excitonic lineshape and the parasitic background absorption can be found in [46].

It should be noted that samples used in recent cooling experiments have a $\tau_{nr} \simeq$ 27 µs [18]. According to our analysis, it should be possible to see laser cooling with these samples. However net cooling has not been observed. It is therefore necessary to analyze possible uncertainties of the parameter values that we have used in the preceding analysis. As an example, we show in Figure 6.10 the cooling results with various parameter values for the background absorption. We can see that, as expected, the cooling threshold increases with increasing $\alpha_b$. We can also see that, for a value of $\alpha_b = 64$ cm$^{-1}$, cooling becomes impossible for temperatures below 30 K. Further studies of parameter variations for the Auger coefficient and the extraction coefficient have been presented in [46]. That study leads us to believe that a combination of slight parameter variations may explain the fact that net cooling of bulk GaAs has not been observed yet.

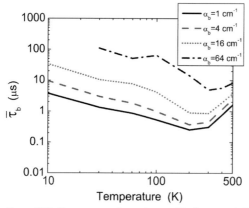

**Figure 6.10** Temperature dependence of $\bar{\tau}_b$ for several different $\alpha_b$ values using the second-order model. From [52].

## 6.5 Cooling of GaAs Quantum Wells

Until now, our cooling analysis has been directed at bulk samples. However, in efficiency terms, semiconductor quantum wells have some potential advantages that are derived from their broken translational invariance, such as fast near-equilibrium temperature equilibration by phonon scattering [12, 50, 51]. In order to study quantum wells using the general theoretical framework discussed in previous sections [52], we need to use electron and hole Green's functions for a quasi-two-dimensional system. We also need to generalize the KMS relation. In contrast to the electrons and holes, the light is not confined to the quasi-two-dimensional space of the quantum well. One cannot formulate the concept of an absorption coefficient. Instead, one usually defines the normal-incidence absorbance of the quantum well, which can be shown via a transfer matrix approach to be

$$A(\omega) \equiv 1 - \frac{I_r}{I_0} - \frac{I_t}{I_0} \approx \operatorname{Im} \sigma^R(\omega), \quad (6.25)$$

where

$$\sigma^R(\omega) = 4\pi \frac{\omega}{n_b c} \chi^R(\omega) \quad (6.26)$$

and the incident, transmitted and reflected intensities are denoted by $I_0$, $I_t$ and $I_r$, respectively, while $n_b$ is the refractive index of the barrier material. The quantum well's susceptibility is obtained through the same diagrammatic analysis as used in the previous section (see Figure 6.1), where all wavevectors are now two-dimensional in-plane wave vectors, and the Coulomb potential is $V_q = (2\pi e^2)/(\mathcal{A}\varepsilon_b q)$, where $\mathcal{A}$ is the area of the quantum well. The generalized KMS relation for the transition between the conduction band and the heavy-hole valence band was derived in [20]:

$$R(\omega) = \frac{2}{3} \left( \frac{\omega n_b}{\pi c} \right)^2 \left( \frac{1}{e^{(\hbar\omega - \mu)/(k_B T)} - 1} \right) \operatorname{Im} \sigma^R(\omega), \quad (6.27)$$

where $\mu$ is the total chemical potential of the electron–hole pair. The factor 2/3 is a consequence of the selection rules for this transition (see [53, 54]).

For the cooling analysis, we also need information about nonradiative recombination (see for example [55, 56]) and Auger recombination rates [57, 58] in quantum wells. In general, the temperature dependence of the Auger coefficient is a complicated function of the exact quantum well configuration. In terms of experimentally verified theories that describe Auger recombination at low temperatures, the situation in quantum wells seems to be even worse than in bulk. It it difficult to find theories with experimental verification of the Auger coefficients of quantum wells at low temperatures. A simple model for quantum wells of thickness $L > 50$ nm, which is consistent with the literature and sufficient for our cooling analysis, is

$$C^{(2D)} = 10^{17} \left( \frac{10 \text{ nm}}{L} \right)^2 \text{cm}^4 \text{s}^{-1}. \quad (6.28)$$

Of course, the applicability of our luminescence theory depends crucially on how close the experimental system is to quasi-equilibrium. In general, if the pump-generated electron–hole pairs are thermalized within the bands much faster than the loss of e–h pairs due to radiative or nonradiative transitions, one can assume the plasma or partially ionized exciton gas to be close to quasi-thermal equilibrium (provided the pump intensity is not too high). For bulk semiconductors under cw excitation, this condition seems to be satisfied. In particular, the experimental luminescence spectra shown in [38] are consistent with the assumption of quasi-thermal equilibrium. So do the luminescence spectra for quantum wells shown in that work. However, other works [59–63] have provided evidence that, at very low temperatures, quasi-thermal equilibrium is not reached by the excited carriers in thin quantum wells before they decay. So the results shown in the following must be viewed with this limitation in mind.

As in the analysis for bulk GaAs, we define a radiative recombination coefficient, $B(N, T) = (1/N^2) \int d\omega R(\omega, N, T)$, for use in the cooling analysis. This coefficient is shown in Figure 6.11 as a function of pair density for various temperatures. Its general behavior is similar to that seen for bulk semiconductors (see Figure 6.5). Also, the optimal cooling densities, indicated in Figure 6.11, are again relatively high (generally less than but close to the Mott density).

In Figure 6.12 we show our calculated minimum breakeven nonradiative lifetime as a function of temperature using four approximations for calculating the luminescence, which were defined in the last section. The parasitic absorption per quantum well is set at $\sigma_b = 1.235 \times 10^{-6}$. A stack of 81 quantum wells with this level of $\sigma_b$ would give a total parasitic absorption roughly equal to that we used in the bulk analysis. The extraction coefficient is set at $\eta_e = 0.25$. Effects of light propagation and reabsorption inside the multiple quantum well structure are not explicitly considered. This limits the applicability of the model to small quantum well stacks.

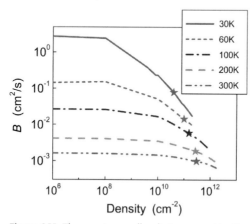

**Figure 6.11** The quantum well radiative recombination coefficient $B$ as a function of electron–hole density at various temperatures. The star on each curve marks the optimal density for cooling. From [52].

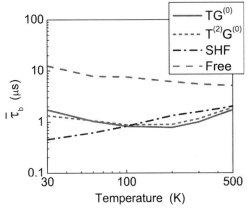

**Figure 6.12** Break-even nonradiative lifetime required for cooling in a quantum well system as a function of temperature calculated at various levels of approximation. *Free*: free fermion plasma. *SHF*: Ladder approximation for the susceptibility with Hartree–Fock and Coulomb hole self-energy. $T^{(2)}G^{(0)}$: SHF + second-order self-energy. $TG^{(0)}$: self-consistent $T$ matrix self-energy. See text for more detail. For each curve, cooling is possible if the actual sample nonradiative lifetime is above the curve. From [52].

As in bulk GaAs, the excitonic enhancement of absorption/luminescence at low frequencies lowers $\bar{\tau}_b$ and hence improves the prospects of cooling. The difference in the results of $TG^{(0)}$ and $T^{(2)}G^{(0)}$ gives an indication of the effects of including the appropriate vertex corrections in the $TG^{(0)}$ calculations. These two curves give similar results, indicating that adding vertex corrections will probably not change the results.

We note two differences between quantum well and bulk GaAs that are relevant to the behavior of $\bar{\tau}_b$.

(i) At equivalent levels of background parasitic absorption, excitonic correlations are more important in enabling cooling at low temperatures in bulk than in quantum wells. This can be understood as follows. In order to achieve cooling, the semiconductor must be pumped below the mean luminescence, and the absorption at the pump frequency must be significantly higher then the background absorption. For a free Fermi gas, the 2D density of states is much higher than its 3D counterpart at the low-frequency end of the absorption/luminescence spectrum. This allows the 2D free model to have sufficient absorption at the pump frequency, whereas in 3D excitonic effects are needed to provide sufficient absorption. Hence, in the free model of Figure 6.9, cooling was shown to be impossible for bulk GaAs below 30 K at the assumed level of background absorption, whereas in quantum wells, excitonic effects serve merely to improve $\bar{\tau}_b$ by an order of magnitude in Figure 6.12.

(ii) While the Auger coefficient in bulk GaAs decreases exponentially with temperature, the quantum well Auger coefficient varies much more slowly [57, 58]. As a result, Auger effects are negligible in the cooling analysis in bulk GaAs below 100 K, whereas in quantum wells, they are still appreciable in the determination of $\bar{\tau}_b$ at lower temperatures.

Available experimental measurements of the nonradiative lifetime in GaAs-based QWs [55, 56] give values of between 1 and 10 ns, which is two orders of magnitude shorter than the $\bar{\tau}_b$ shown in Figure 6.12. While in bulk semiconductors the nonradiative lifetime increases exponentially with decreasing temperature, [55] reports a decrease below approximately 100 K, with a maximum nonradiative lifetime of about 10 ns. Reference [56] reports nonradiative lifetimes in shallow GaAs quantum wells that do increase with decreasing temperature below 100 K, but the maximum nonradiative lifetimes are still on the order of 10 ns. Unless this nonradiative lifetime can be improved dramatically, laser cooling may be difficult to achieve in quantum wells. This contrasts with the rather favorable estimation for bulk GaAs.

## 6.6
## Cooling of Doped Bulk Semiconductors

In this section, we further extend our studies and make our theory more applicable to actual samples used in ongoing experiments. Those samples are typically doped. The doping is often not intentional, but a by-product of complex sample configurations which are heterostructures that include surface layers aimed at reducing nonradiative surface recombination. The bulk GaAs layers sandwiched between the surface layers are usually p-doped. It is therefore important to account for the additional luminescence channel opened by the presence of acceptors in our microscopic theory. We focus on shallow impurities with acceptor binding energies comparable to the bulk exciton binding energy. Specifically, we will study the influence of carbon impurities with an acceptor binding energy of 26.5 meV. Due to the strong on-site Coulomb repulsion, the acceptor population is (see for example [64]):

$$f_a(\varepsilon_A) = \frac{1}{e^{\beta(\varepsilon_A - \mu_h)} + 2d} \, . \tag{6.29}$$

These statistics differ from those of noninteracting carriers by the appearance of the term $2d$ (the factor of two accounts for the spin and $d$ accounts for the orbital degeneracy of the acceptor level), which ensures that an acceptor site can have at most one hole. In [65] we introduced a modified Green's function appropriate for the acceptor impurity states:

$$G_a^<(m, \omega) = i f_h(\omega) \widetilde{A}_a(m, \omega) \tag{6.30}$$

$$G_a^>(m, \omega) = -i(1 - f_h(\omega)) \widetilde{A}_a(m, \omega) \tag{6.31}$$

$$\widetilde{A}_a(m, \omega) = \frac{1 + e^{\beta(\hbar\omega - \mu_h)}}{2d + e^{\beta(\hbar\omega - \mu_h)}} 2\pi\hbar\delta(\hbar\omega - \varepsilon_A) \, . \tag{6.32}$$

We note that in this case $\widetilde{A}_a$ does not necessarily carry the meaning of a spectral function. The acceptor Green's function $G_a$ can be included in the expression for the optical susceptibility, as shown in Figure 6.1c. We assume that phonon and Coulomb interactions allow the acceptor sites and valence band to interact sufficiently to thermalize before carriers recombine, thus allowing us to use the same chemical potential for both the acceptor site and valence band.

The cooling theory has to be modified as follows. The net power is now given by

$$P_{net} = I[a(\omega_a, n_e) + a_b] - \hbar\omega_l \eta_e B(n_e) n_e (n_e + N_a) . \quad (6.33)$$

In the steady state, the density is given by

$$\frac{a(\omega_a, n)}{\hbar\omega_a} I = An + \eta_e B(n_e) n_e (n_e + N_a) + C n_e n_h^2 . \quad (6.34)$$

Here, $n_e$ is the optically generated electron carrier density,

$$n_h = n_e + n_a , \quad (6.35)$$

is the density of holes in the valence band,

$$n_a = N_a[1 - 2df_a(\varepsilon_A)] , \quad (6.36)$$

which is the density of holes in the valence band that come from the acceptors, and $B(n_e)$ is the radiative recombination coefficient. The extracted luminescence, $\eta_e B(n_e) n_e (n_e + N_a)$ is calculated from the microscopic theory analogously to the case of undoped bulk GaAs.

The optimal electron and hole densities are shown in Figure 6.13. The general trends are similar to the case of the undoped semiconductor (see Figure 6.8). At high temperatures, the optimum density decreases mainly because of the increasing Auger recombination. The decrease in the optimum density at low temperatures is dominated by phase-space filling effects. In contrast to the undoped case, the electron and hole densities are generally not equal in Figure 6.13. With a doping density of $N_a = 3 \times 10^{16}$ cm$^{-3}$, $n_e$ and $n_h$ can differ at most by this amount. However, at low temperatures, the number of valence-band holes generated by acceptors becomes very small, and so $n_e$ and $n_h$ become equal in this limit.

In Figure 6.14 we show the optimal break-even nonradiative lifetime. Here, we compare the case of an undoped semiconductor with that of the p-doped semiconductor with a relatively small doping density of $N_a = 3 \times 10^{16}$ cm$^{-3}$. We see that both cases show the general features already discussed above (Figure 6.9). Only at very low temperatures does the doped case exhibit a slightly less favorable cooling threshold. This may be attributed to the onset of band-to-acceptor luminescence,

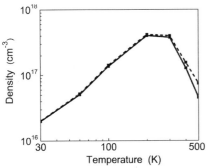

**Figure 6.13** Optimal cooling density vs. temperature. The solid line shows the electron density and the dashed line the hole density. The doping density is $N_a = 3 \times 10^{16}$ cm$^{-3}$. From [65].

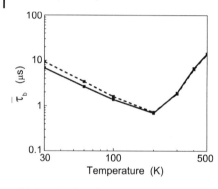

**Figure 6.14** Break-even nonradiative lifetime vs. temperature for several different theories. For each model, the predicted cooling region is above its curve. The *solid line* shows the result for the undoped structure, and the *dashed line* corresponds to a doping density of $N_a = 3 \times 10^{16}$ cm$^{-3}$. From [65].

which emits low-frequency photons (lower than the optimal absorption frequency), and therefore impedes cooling. At the doping density considered here, this impediment is not very significant, but ongoing studies (to be published in the future) indicate that higher doping densities can lead to a more drastic deterioration of the cooling threshold.

We finally note that our analysis of the influence of doping is restricted to the modification of the luminescence spectra. If the presence of acceptors leads to a significant reduction in the nonradiative lifetime, the results shown in Figure 6.14 have to be interpreted in the context of that reduction. While the thresholds of the doped and undoped case shown in Figure 6.14 are very similar, a reduction of the lifetime in the doped case may make it more difficult to reach the threshold.

## 6.7
## Conclusion

In conclusion, we have developed a comprehensive theory for semiconductor laser cooling based on a microscopic theory for a partially ionized exciton gas in semiconductors, including doped and undoped bulk semiconductors as well as semiconductor quantum wells. The theory, formulated in terms of a real-time nonequilibrium Keldysh Green's function technique, evaluates the single-particle and two-particle Green's function at the self-consistent T-matrix level, with a quasi-static plasmon-pole screening model. The theory is applicable to systems in quasi-thermal equilibrium at temperatures ranging from the few Kelvins regime to above room temperature. The results for luminescence and absorption spectra have been used in a cooling theory that generalized the rate equation approach of Sheik-Bahae and Epstein. We have evaluated the cooling criterion expressed as a break-even condition for the nonradiative lifetime required for cooling ($\bar{\tau}_b$), and we found that it is strongly affected by excitonic effects. In bulk semiconductors at low temperatures, where Auger recombination can be neglected, these effects lead to a dramatic reduction in $\bar{\tau}_b$ compared to the case where only phase-space filling is taken into account. An important aspect of exciton effects is the large enhancement of the (density-dependent) radiative recombination rate, especially at low temperatures. We found that the exciton resonance helps overcome parasitic background absorption effects, and quite generally that excitonic effects are crucial for cooling at low temperatures. On the other hand, excitonic effects do not significantly influence the optimal upconversion (i.e. the upconversion evaluated with the optimal absorption frequency). Light propagation effects, such as luminescence reabsorption and surface reflection, have been modeled by the extraction efficiency parameter. Initial results from our ongoing investigations, not described in detail in this chapter, indicate that an accurate fully microscopic theory involving the photon Green's function is possible. Results of that ongoing research, which also include the almost always negligible effects of polaritons, will be published elsewhere. We extended the investigation to semiconductor quantum wells as well as doped bulk semiconductors. Under the present assumption of quasi-thermal equilibrium, we did not find any indications that quantum wells are better suited to optical refrigeration than bulk semiconductors. As for the influence of impurities (here: acceptors), we found that the additional low-frequency luminescence and absorption channel (here: between the conduction band and the acceptor states) does not significantly deteriorate the cooling threshold $\bar{\tau}_b$, as long as the impurity concentration is not too high (here: $3 \times 10^{16}$ cm$^{-3}$). Ongoing studies not presented in this chapter indicate that higher doping concentrations lead to an increase in $\bar{\tau}_b$, or even make cooling impossible (at sufficiently low temperatures). We have not addressed the issue of impurity-induced nonradiative recombination, which may adversely affect cooling, even in cases where $\bar{\tau}_b$ is not affected.

In the future, it would be desirable to generalize our theory to a fully nonequilibrium and possibly time-dependent analysis, to study bandstructure effects beyond the parabolic two-band model, to include charge carrier and light transport effects in realistic heterostructure geometries, and to use microscopic theories for all processes that enter into the cooling theory (including Auger recombination and nonradiative recombination). This would further broaden the predictive power of the theory.

## Acknowledgement

We thank M. Sheik-Bahae, B. Imangholi, M. Hasselbeck, R. Epstein and J. Khurgin for helpful discussions. We acknowledge financial support from AFOSR (MURI Grant #FA9550-04-1-0356), and additional support from JSOP.

## References

1 PRINGSHEIM, P. (1929) *Z. Phys.*, 57, 739.
2 EPSTEIN, R.I., BUCHWALD, M.L., EDWARDS, B.C., GOSNELL, T.R. AND MUNGAN, C.E. (1995) *Nature*, 377, 500.
3 EDWARDS, B.C., ANDERSON, J.E., EPSTEIN, R.L., MILLS, G.L. AND MORD, A.J. (1999) *J. Appl. Phys.*, 86, 6489.
4 GOSNELL, T.R. (1999) *Opt. Lett.*, 24, 1041.

5 Hoyt, C.W., M. Sheik-Bahae, Epstein, R.I., Edwards, B.C. and Anderson, J.E. (2000) *Phys. Rev. Lett.*, **85**, 3600.
6 Thiede, J., Distel, J., Greenfield, S.R. and Epstein, R.I. (2005) *Appl. Phys. Lett.*, **86**, 154107.
7 Fernandez, J., Garcia-Adeva, A.J. and Balda, R. (2006) *Phys. Rev. Lett.*, **97**, 033001.
8 Clark, J.L. and Rumbles, G. (1996) *Phys. Rev. Lett.*, **76**, 2037.
9 Oraevsky, A.N. (1996) *J. Russ. Laser Res.*, **17**, 471.
10 Gauck, H., Gfroerer, T.H., Renn, M.J., Cornell, E.A. and Bertness, K.A. (1997) *Appl. Phys. A*, **64**, 143.
11 Rivlin, L.A. and Zadernovsky, A.A.(1997) *Opt. Commun.*, **139**, 219.
12 Sheik-Bahae, M. and Epstein, R.I. (2004) *Phys. Rev. Lett.*, **92**, 247403.
13 Danhong Huang, Apostolova, T., Alsing, P.M. and Cardimona, D.A. (2004) *Phys. Rev. B*, **70**, 033203.
14 Apostolova, T., D. Huang, Alsing, P.M. and Cardimona, D.A. (2005) *Phys. Rev. A*, **71**, 013810.
15 Li, J. (2007) *Phys. Rev. B*, **75**, 155315.
16 Khurgin, J.B. (2007) *Phys. Rev. Lett.*, **98**, 177401.
17 Khurgin, J.B., Sun, G. and Soref, R.A. (2007) *J. Opt. Soc. Am. B*, **24**, 1968.
18 Imangholi, B., Hasselbeck, M.P., Sheik-Bahae, M., Epstein, R.I. and Kurtz, S. (2005) *Appl. Phys. Lett.* **86**, 081104.
19 Zimmermann, R. (1987) *Many-Particle Theory of Highly Excited Semiconductors*, Teubner, Leipzig.
20 Kwong, N.H., Rupper, G., Gu, B. and Binder, R. (2007) *Proc. SPIE*, **6461**, 6461OI1.
21 Kadanoff, L.P. and Baym, G. (1989) *Quantum Statistical Mechanics*, Addison-Wesley, New York.
22 Haug, H. and Jauho, A.P. (1996) *Quantum Kinetics in Transport and Optics of Semiconductors*, Springer, Berlin.
23 Schäfer, W. and Wegener, M. (2002) *Semiconductor Optics and Transport – From Fundamentals to Current Topics*, Springer, Berlin.
24 Danielewicz, P. (1984) *Ann. Phys.*, **152**, 239.
25 Löwenau, J.P., Reich, F.M., and Gornik, E. (1995) *Phys. Rev. B*, **51**, 4159.
26 Schmielau, T., Manske, G., Tamme, D. and Henneberger, K. (2000) *Phys. Stat. Sol. B* **221**, 215.
27 Tassone, F. and Piermarocchi, C. (1999) *Phys. Rev. Lett.*, **82**, 843.
28 Piermarocchi, C. and Tassone, F. (2001) *Phys. Rev. B*, **63**, 245308.
29 Haug, H. and Schmitt-Rink, S. (1984) *Prog. Quantum Electron.*, **9**, 3.
30 Schäfer, W., Binder, R. and Schuldt, K.H. (1988) *Z. Physik B*, **70**, 145.
31 Schäfer, W., Lövenich, R., Fromer, N.A. and Chemla, D.S. (2001) *Phys. Rev. Lett.*, **86**, 344.
32 Rupper, G., Kwong, N.H. and Binder, R. (2009), to be published
33 Huang, K. (1987) *Statistical Mechanics*, John Wiley & Sons, Inc., New York.
34 Zimmermann, R. and Stolz, H. (1985) *Phys. Stat. Sol. B*, **131**, 151.
35 Portnoi, M.E. and Galbraith, I. (1999) *Phys. Rev. B*, **60**, 5570.
36 Siggelkow, S., Hoyer, W., Kira, M. and Koch, S.W. (2004) *Phys. Rev. B*, **69**, 073104.
37 Gartner, P., Banyai, L. and Haug, H. (2000) *Phys. Rev. B*, **62**, 7116.
38 Venu Gopal, A., Kumar, R., Vengurlekar, A.S., Bosacchi, A., Franchi, S. and Pfeiffer, L.N. (2000) *J. Appl. Phys.*, **87**, 1858.
39 Liebler, J.G., Schmitt-Rink, S. and Haug, H. (1985) *J. Luminescence*, **34**, 1.
40 Liebler, J. and Haug, H. (1991) *Europhys. Lett.*, **14**, 71.
41 von Lehmen, A., Zucker, J.E., Heritage, J.P. and Chemla, D.S. (1987) *Phys. Rev. B*, **35**, 6479.
42 Bonnot, A. and ala Guillaume, C.B. (1974) In *Polaritons*, (eds E. Burstein and F. de Martini), Pergamon, New York, pp. 197–202.
43 Khurgin, J. (2006) *Proc. SPIE*, **6115**, 611519.
44 Hader, J., Moloney, J.V. and Koch, S.W. (2005) *Appl. Phys. Lett.*, **87**, 201112 (3).
45 Wang, F., Wu, Y., Hybertson, M.S. and Heinz, T.F. (2006) *Phys. Rev. B*, **73**, 245424 (5).

46 Rupper, G., Kwong, N.H. and Binder, R. (2007) *Phys. Rev. B*, **76**, 245203.
47 Zimmermann, R., Kilimann, K., Kraeft, W.D., Kremp, D. and Röpke, G. (1978) *Phys. Stat. Sol. B*, **90**, 175.
48 Haug, H. and Koch, S.W. (2004) *Quantum Theory of the Optical and Electronic Properties of Semiconductors*, 4th ed., World Scientific, Singapore.
49 Sheik-Bahae, M., Imangholi, B., Hasselbeck, M.P., Epstein, R.I. and Kurtz, S. (2006) *Proc. SPIE*, **6115**, 611518.
50 Ivanov, A.L., Littlewood, P.B. and Haug, H. (1999) *Phys. Rev. B*, **59**, 5032.
51 Butov, L.V., Lal, C.W., Gossard, A.C. and Chemla, D.S. (2002) *Nature*, **417**, 47.
52 Rupper, G., Kwong, N.H., Gu, B. and Binder, R. (2008) *Phys. Stat. Sol. B*, **245**, 1049.
53 Andreani, L.C. and Pasquarello, A. (1990) *Phys. Rev. B*, **42**, 8928.
54 Runge, E. (2002) *Solid State Physics vol. 57*, Elsevier, Amsterdam, pp. 149–305.
55 Gurioli, M., Vinattieri, A., Colocci, M., Deparis, C., Massies, J., Neu, G., Bosacchi, A. and Franchi, S. (1991) *Phys. Rev. B*, **44**, 3115.
56 Tignon, J., Heller, O., Roussignol, P., J. Martinez-Pastor, LeLong, P., Bastard, G., Iotti, R.C., Andreani, L.C., Thierry-Mieg, V. and Planel, R. (1998) *Phys. Rev. B*, **58**, 7076.
57 Hausser, S., Fuchs, G., Hangleiter, A. and Streubel, K. (1990) *Appl. Phys. Lett.*, **56**, 913.
58 Polkovnikov, A.S. and Zegrya, G.G. (1998) *Phys. Rev. B*, **58**, 4039.
59 Schnabel, R.F., Zimmermann, R., Bimberg, D., Nickel, H., Losch, R. and Schlapp, W. (1992) *Phys. Rev. B*, **46**, 9873.
60 Piermarocchi, C., Tassone, F., Savona, V., Quattropani, A. and Schwendimann, P. (1996) *Phys. Rev. B*, **53**, 15834.
61 Kira, M., Hoyer, W., Stroucken, T. and Koch, S.W. (2001) *Phys. Rev. Lett.*, **87**, 176401.
62 Hoyer, W., Kira, M. and Koch, S.W. (2003) *Phys. Rev. B*, **67**, 155113.
63 Chatterjee, S., Ell, C., Mosor, S., Khitrova, G., Gibbs, H.M., Hoyer, W., Kira, M., Koch, S.W., Prineas, J.P. and Stolz, H. (2004) *Phys. Rev. Lett.*, **92**, 067402.
64 Ashcroft, N.W. and Mermin, N.D. (1976) *Solid State Physics*, Sounders College, New York.
65 Rupper, G., Kwong, N.H., Gu, B. and Binder, R. (2008) *Proc SPIE*, **6907**, 690705-1.

# 7
# Improving the Efficiency of Laser Cooling of Semiconductors by Means of Bandgap Engineering in Electronic and Photonic Domains
*Jacob B. Khurgin*

## 7.1
## Introduction

Preceding chapters of this book provide a relatively comprehensive picture of the state of the art of laser refrigeration in solids. The authors of these chapters describe both successes attained on the way to practical implementations of laser cooling of solids and the formidable challenges that remain on this path. While the first media in which laser refrigeration was successfully demonstrated were the rare-earth doped glasses and crystals [1–4], from a practical point of view it is far more preferable to achieve laser cooling in a semiconductor because such a cooler can be easily integrated with electronic and optical devices. This fact has not been lost on researchers, and for the last few years the feasibility of laser cooling in semiconductors has been studied extensively [5–8]. The difficulties involved in achieving laser cooling in semiconductors are well understood – they are a relatively high degree of nonradiative recombination, substantial background absorption, and also the low extraction efficiency engendered by the high value of the refractive index.

Consider the schematics of laser cooling, as shown in Figure 7.1. It is easy to see that for each laser photon of energy $h\nu_L$ absorbed near the band gap and then re-emitted at a somewhat higher energy $h\nu_F$, the net cooling can be on the order of $kT$. At the same time, if the absorbed photon is not re-emitted but instead transfers its energy to the lattice, net heating of roughly $h\nu_L$ ensues. Based on this principle, it is not difficult to understand that the total radiative efficiency of the emission should exceed $1 - k_B T / h\nu_L$ for the net cooling to take place.

To quantify these considerations, one can introduce the cooling efficiency as the ratio of cooling power to the laser power as

$$\eta_{\text{cool}} = \frac{P_{\text{cool}}}{P_{\text{in}}} \approx \frac{\tau_{\text{nr,b}}^{-1} - \tau_{\text{nr}}^{-1}}{\eta_e \tau_{\text{rad}}^{-1}}, \tag{7.1}$$

where $\eta_e$ is the light extraction efficiency, $\tau_{\text{nr}}^{-1}$ is the rate of nonradiative recombination, which does not depend on carrier concentration (typically Shockley–Reed–Hall recombination on traps and surface recombination), and $\tau_{\text{nr,b}}^{-1}$ is the critical

*Optical Refrigeration. Science and Applications of Laser Cooling of Solids.*
Edited by Richard Epstein and Mansoor Sheik-Bahae
Copyright © 2009 WILEY-VCH Verlag GmbH & Co. KGaA, Weinheim
ISBN: 978-3-527-40876-4

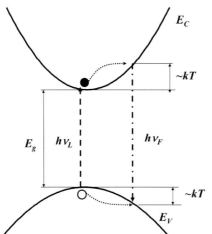

Figure 7.1 Principle of laser cooling of semiconductors.

(break-even) nonradiative recombination rate:

$$\tau_{nr,b}^{-1} = \eta_e \left[ \frac{\langle \nu_F \rangle / \nu_L}{1 + \alpha_b/\alpha(\nu_L)} - 1 \right] \tau_{rad}^{-1} - C(T) N_e^2 , \qquad (7.2)$$

where $\alpha_b$ is the background absorption, $N_e$ is the carrier concentration, and $C(T)$ is the Auger recombination coefficient. The mean fluorescence energy can be found as

$$\langle h\nu_F \rangle = \int h\nu_F R(h\nu_F) \, dh\nu_F \Big/ \int R(h\nu_F) \, dh\nu_F , \qquad (7.3)$$

where $R(h\nu_F)$ is the radiative decay rate, which is related to the radiative decay time $\tau_{rad}$ as

$$\int R(h\nu_F) \, dh\nu_F = N_e \tau_{rad}^{-1} . \qquad (7.4)$$

Thus, $\tau_{nr,b}$ is a good figure of merit for a given laser refrigeration scheme. This break-even time had been estimated to be in the range of 0.1 to 10 μs [8] for different materials and temperatures, which puts very strict requirements on the quality of the material. Thus, work in laser refrigeration should be geared towards reducing $\tau_{nr,b}$. According to (7.2), it is desirable to increase the "blueshift" of the mean fluorescence relative to the laser photon energy $\langle \nu_F \rangle - \nu_L$ while maintaining a fast overall recombination rate and keeping absorption at the frequency of the laser high. Clearly, if one can modify the radiative rate to have a step-like character (i.e. reasonably low near the laser frequency and high a few $kT$ above it), then one can hope to greatly increase the efficiency, since only the carriers with relatively high energies will be able to recombine and the average emission frequency will increase (as shown in Figure 7.2). There are two ways of achieving the "step-like" spectrum. The preferable method would be to enhance the radiative recombination rate at higher energies, but enhancing the radiative rate is not a trivial feat, as will become

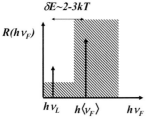

Figure 7.2 "Ideal" recombination rate for laser cooling.

clear later on in this chapter. Therefore, one is typically forced to use the other alternative and simply try to block the radiative recombination at lower energies. However, by blocking the transitions, the overall rate of radiative emission $\tau_{rad}^{-1}$ is also reduced, and – potentially more damaging – the absorption at the frequency of the laser also diminishes, with negative consequences. Therefore, the engineering of transition rates involves studying trade-offs and compromises.

The radiative transition rate at a given photon energy $h\nu_F$ can be written as a product

$$R(h\nu_F) \sim M^2(h\nu_F)\varrho_{cv}(h\nu_F)\varrho_{ph}(h\nu_F) , \qquad (7.5)$$

where $M$ is the matrix element of the radiative transition, $\varrho_{cv}$ is the reduced density of electron–hole states, and $\varrho_{ph}$ is the density of photon states. From this, it is evident that – like many other processes – laser cooling is a "density-of-states game", and so one can affect the radiative transition rate by modifying the density of states. In fact, all three components of $R$ can be modified:

1. The density of electron–hole states $\varrho_{cv}$ can be modified using QWs or simply impurity bands. This method is explored in Section 7.2. One could also use of an Urbach tail caused by phonon-assisted transitions as a density of states that is "custom tailored" for laser cooling. We investigate this method in Section 7.3 of this chapter.

2. The matrix element of transition $M$ (oscillator strength) can be altered by using heterostructures that combine direct and indirect transitions in real space; for instance, type II QWs can be used. Section 7.4 is dedicated to this method.

3. The density of photonic states $\varrho_{ph}$ can be modified in photonic bandgap structures or using surface plasmon polaritons. This method is considered in Sections 7.5 and 7.6.

## 7.2
### Engineering the Density of States Using Donor–Acceptor Transitions

Let us now investigate the possibility of modifying the density of states using donor–acceptor pairs to improve the quantum efficiency of laser cooling. This idea, shown in Figure 7.3, is very simple – basically we want absorption to take place from acceptor to donor state [9] while radiative recombination takes place between the band edges (i.e. at an energy that is larger than the energy of the absorbed photon by the sum of the binding energies of the donor and acceptor, $E_D + E_A$). Then we would expect to see an enhancement of the quantum efficiency from roughly $kT/h\nu_L$ to roughly $(kT + E_D + E_A)/h\nu_L$, as predicted in [10].

The main challenge of this method is to ensure that most of the photogenerated carriers get ionized into the bands while the absorption is sufficient. Therefore the doping density and the binding energies of the donors and acceptors must be carefully selected. We have performed detailed calculations for compensation-doped GaAs at different temperatures. The results are shown in Figures 7.4 and 7.5. The horizontal axis on all plots is the ionization energy of the acceptor; the ionization energy of the donor is scaled by the density of states ratio, $E_D = (m_c/m_v)^{3/2} E_A$, to ensure an equal probability of ionization. The doping density was taken to be $N_D = N_A = 10^{15}$ cm$^{-3}$ to ensure that the absorption at the laser wavelength is sufficiently high, $\alpha(\nu_L) \sim 10$ cm$^{-1}$. Then the background absorption $\alpha_b \sim 1$ cm$^{-1}$ in (7.2) can be neglected. At the same time, the doping density is sufficiently small to ensure that most of the carriers stay ionized with an absorption coefficient on the order of 10 cm$^{-1}$.

Let us first turn our attention to Figure 7.4 for GaAs at 100 K. As one expects, the quantum efficiency, shown in normalized units as

$$\eta'_Q = \left[\langle \nu_F \rangle / \nu_L - 1\right] E_{gap}/(kT) \tag{7.6}$$

in Figure 7.4a, initially increases as the binding energy increases, since each electron–hole pair must acquire additional energy from the lattice to become ionized. However, as the binding energy increases further, the electrons and holes find it more and more difficult to become ionized; donor–acceptor pair recombination becomes the dominant mechanism, and the quantum efficiency quickly falls off. Unfortunately, as the binding energy increases the donor–acceptor level population quickly becomes degenerate at a relatively low value of carrier density $N_{e,sat}$, as shown in Figure 7.4b. When the population is degenerate, the radiative recombination no longer has a bimolecular character $\tau_{rad}^{-1} \sim N_e$ but instead becomes linear (i.e. is characterized by constant $\tau_{rad}^{-1}$). Thus, we cannot decrease the break-even recombination rate beyond a certain point no matter how hard

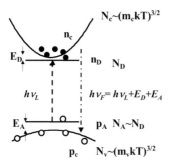

Figure 7.3 Principle of density-of-state engineering for laser cooling using doping states.

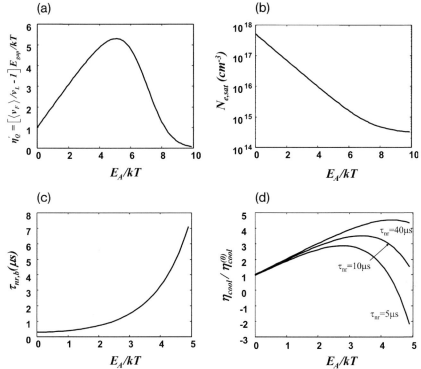

**Figure 7.4** Dependence of various cooling characteristics on the acceptor binding energy for GaAs at 100 K with doping density $N_D = N_A = 10^{15}$ cm$^{-3}$. (a) Quantum efficiency, (b) saturation density, (c) critical nonradiative recombination time for $\eta_E = 50\%$, and (d) enhancement of the cooling efficiency for various nonradiative recombination times.

we pump the semiconductor. In addition, as the donor–acceptor levels become occupied at degeneracy, the absorption itself saturates. As a result, the break-even nonradiative recombination time (Figure 7.4c) always increases with the ionization energy. However, lifetimes as long as tens of microseconds have recently been achieved in GaInP/GaAs heterostructures [11]. Therefore, the fact that the nonradiative time increases should not be interpreted as a major impediment to laser refrigeration.

Figure 7.4d plots the cooling efficiency normalized to the cooling efficiency of undoped GaAs $\eta_{cool}^{(0)}$ for various realistic values of nonradiative lifetimes. It is obvious that, provided a sufficiently long nonradiative recombination time can be attained, the cooling efficiency can indeed be enhanced by as much as a factor of five using a compensation-doped structure.

Since only this figure showing the cooling efficiency enhancement is of practical interest, only this dependence is shown for the 300 K case in Figure 7.5. One can see from this figure that an enhancement of the laser cooling is again possible, but that it is not as strong as the one seen at 100 K.

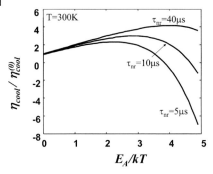

**Figure 7.5** Enhancement of the cooling efficiency for various nonradiative recombination times vs. the acceptor binding energy for GaAs at 300 K. $N_D = N_A = 10^{15}$ cm$^{-3}$, $\eta_E = 50\%$.

## 7.3
## Refrigeration Using Phonon-Assisted Transitions

As shown in the previous section, the deviation of the density of states in the vicinity of the bandgap from the simple parabolic law can be quite beneficial to laser refrigeration. Intentional doping and the induction of below-bandgap transitions is just one way to accomplish this goal. In fact, even in the absence of intentional doping, the simple parabolic band approximation becomes invalid near the bandgap for a variety of reasons, such as impurity states, Coulomb and many-body interactions, and phonon-assisted processes. The roles of excitons and many-body effects have been considered in various works [7, 8], and they have been shown to be important and mainly beneficial to the task of laser cooling. In [10], the role of discrete donor and acceptor states was investigated. However, in real semiconductors the states near the bandgap do not follow the parabolic law but instead exhibit an exponential Urbach tail [12, 13]. The origin of the Urbach tail had been linked to either impurities [14–17], interactions with phonons [18, 19], or both [20].

From the point of view of laser cooling, the simplest way to deal with the Urbach tail is to assume that there are real states below the bandgap energy, as shown in Figure 7.6a. The incoming laser photon then causes a transition between these two impurity-associated states, as shown in Figure 7.6a. Following excitation, thermalization takes place in the band, and the fluorescence photon $\hbar\omega_F$ is finally re-emitted. This model was investigated in [9] and it was shown that in certain circumstances the presence of impurity states can be conducive to laser cooling. However, the presence of impurity states invariably increases the probability of nonradiative decay and also parasitic absorption. It is no wonder then that the best results [21], in which semiconductors were brought very close to the threshold of refrigeration, have been obtained in materials with very low levels of impurities. Therefore, the main reason for the below-the-gap absorption and emission in the samples used for laser cooling is phonon broadening of the absorption.

Typically the Urbach tail can be modeled based on the broadening of absorption and luminescence lines. In this case, the broadening width $\Gamma(\omega)$ is used, which is frequency dependent. The calculation of $\Gamma(\omega)$, usually done using the Green's function approach [22], is quite involved. Therefore, $\Gamma(\omega)$ is usually introduced as

## 7.3 Refrigeration Using Phonon-Assisted Transitions

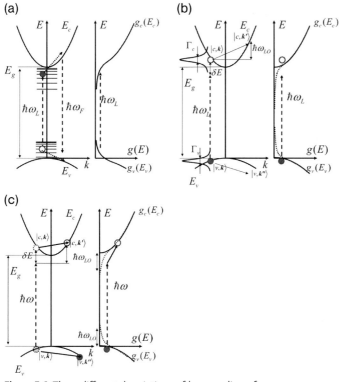

**Figure 7.6** Three different descriptions of laser cooling of a semiconductor near the fundamental bandgap. (a) Bandtail caused by the presence of real states inside the gap. (b) Absorption into the broadened above-the-gap states. Energy is not conserved. (c) Phonon-assisted absorption. Energy is conserved.

a phenomenological parameter, typically as $\Gamma(\omega) \sim \exp\left[a(\hbar\omega - E_g)/k_B T\right]$, where $a$ is chosen to fit experimental data [8]. The problem with the phenomenological model is that it runs into the difficulties when it is necessary to describe the saturation of absorption. This is related to the fact that, strictly speaking, energy is not conserved in the phenomenological model. Indeed, as shown in Figure 7.6b, the transition takes place between the state $|v, k\rangle$ in the valence band (VB) and state $|c, k\rangle$ in the conduction band (CB). These states are broadened due to interactions with the other states in the VB ($|v, k''\rangle$) and in the CB ($|c, k'\rangle$); therefore, in this picture, absorption of the photon with energy $\hbar\omega < E_g$ excites the electron–hole pair with energy $E_{eh}(k) = E_c^k - E_v^k > E_g$, in clear violation of energy conservation. Furthermore, in this model, absorption at frequency $\hbar\omega < E_g$ depends on the carrier population at energies $E_v^k$ and $E_c^k$. Therefore, so long as $E_{eh}$ exceeds the difference between the quasi-Fermi levels of the conduction and valence bands $\Delta\mu_{cv} = \mu_c - \mu_v$ (i.e. the condition $\mu_c - \mu_v < E_c^k - E_v^k$), a photon with energy below $\hbar\omega < E_{eh}$ can be absorbed, even if $\hbar\omega < \Delta\mu_{cv}$. However, this conclusion clearly contradicts the laws of thermodynamics. Thus, to ensure that absorption crosses zero at $\hbar\omega = \Delta\mu_{cv}$, it is necessary to introduce an additional phenomenological dependence of the linewidth on the chemical potential, as was indeed done in [8].

Absorption saturation is a critical concern in laser cooling, because it affects the choice of optimum pump wavelength and carrier concentration, and so it would be highly desirable to avoid all of the uncertainties associated with the phenomenological model. However, as we mentioned above, an exact and explicit treatment of all the phonon-assisted processes is a difficult task, so only the dominant processes should be considered. In the case of laser refrigeration at temperatures above 100 K, the dominant process is a LO-phonon-assisted absorption with only one LO phonon being absorbed. If the excitonic effect is then disregarded, the treatment can be performed within a simple density-matrix framework, and an exact expression for the frequency dependence can be obtained.

To explain the density-matrix treatment [23], we consider four states in the CB with energies $E_c^k$ and $E_c^{k'}$, and two states in the VB with energies $E_v^k$ and $E_v^{k''}$, as shown in Figure 7.6c. The states $|c, k\rangle$ and $|v, k\rangle$ are coupled by the pump optical field of frequency $\omega$ and the coupling Hamiltonian $\hbar\Omega_p = \frac{eP_{cv}A_\omega}{m_0} = \frac{eP_{cv}E_\omega}{m_0\omega}$, where $P_{cv}$ is the momentum matrix element of the interband transition and $A_\omega = \omega^{-1}E_\omega$ is the vector potential of the pump wave. The states $|c, k\rangle$ and $|c, k'\rangle$ are coupled by the Froehlich interaction

$$\hbar\Omega_{cc}^{k,k'} = \left(\frac{e^2 \hbar\omega_{LO}}{2\varepsilon'(k-k')^2 V}\right)^{1/2} n_{LO}, \tag{7.7}$$

where $1/\varepsilon' = 1/\varepsilon_{low} - 1/\varepsilon_\infty$, $\omega_{LO}$ is the LO-phonon frequency, $V$ is the volume, and $n_{LO} = \left(e^{\hbar\omega_{LO}/(k_B T)} - 1\right)^{-1}$ is the occupational number of phonon modes. Similarly, in the valence band, the states $|v, k\rangle$ and $|v, k''\rangle$ are coupled as

$$\hbar\Omega_{vv}^{k,k''} = F_{k,k''} \left(\frac{e^2 \hbar\omega_{LO}}{2\varepsilon'(k-k'')^2 V}\right)^{1/2} n_{LO}, \tag{7.8}$$

where $F_{k,k''}$ is the overlap between the Bloch wave functions of two states in the valence band. Thus, the absorption (and emission) processes are two-step processes involving an LO phonon and a photon, and both energy and momentum are satisfied.

Standard density-matrix analysis leads to an expression for the absorption which is essentially Lorentzian:

$$a(h\nu) = \alpha(E_{gap} + \hbar\omega_{LO})\pi^{-1} n_{LO}$$
$$\times \int \frac{\gamma_c(h\nu, E_{cv}) + \gamma_v(h\nu, E_{cv})}{(E_{cv} - h\nu)^2 + n_{LO}^2 \left[\gamma_c(h\nu, E_{cv}) + \gamma_v(h\nu, E_{cv})\right]^2} \frac{E_{cv}^{1/2}}{E_{gap}^{1/2}} dE_{cv}, \tag{7.9}$$

where $\alpha(E_{cv} + \hbar\omega_{LO})$ is the direct absorption exactly one LO-phonon energy above the absorption edge (equal to $0.45 \times 10^4$ cm$^{-1}$ for GaAs), but with broadenings of the conduction band $\gamma_c(h\nu, E_{cv})$ and valence band $\gamma_v(h\nu, E_{cv})$ that depend on the photon energy; the broadening decreases as the band edge approaches until the broadening (and thus absorption) becomes zero at exactly one LO-phonon energy below the bandgap.

Figure 7.7 shows the results for GaAs at three different temperatures. Near the bandgap at least, the exponential decay of the absorption is apparent, which is no different from the phenomenological model introduced in [8]. The salient feature of (7.8) is that it adequately describes both phonon-assisted absorption below the band edge and direct band-to-band absorption above the band edge, with seamless transition from one to the other at the band edge.

It is also important to evaluate the saturation of the LO phonon below the band-edge absorption. Absorption spectra for different electron–hole pair concentrations are shown in Figure 7.8. It is apparent that the absorption saturates just as if there was a real density of states below the bandgap and the quasi-Fermi level had moved into this absorption edge.

Figure 7.9 shows the total photoluminescence obtained upon combining LO-assisted emission below the band edge and direct emission above the band edge. The shape of the emission curve is close to the desired one shown in Figure 7.2. Note that the average luminescence energy $\hbar \langle \omega_F \rangle$ does not change much with electron density.

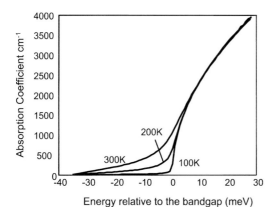

**Figure 7.7** Absorption spectra of GaAs near the band edge at different temperatures.

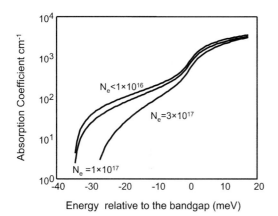

**Figure 7.8** Saturation of GaAs absorption at $T = 200$ K at three different carrier densities.

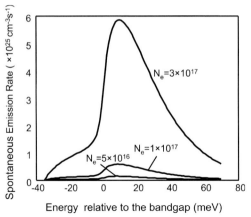

**Figure 7.9** Photoluminescence of GaAs at 200 K at three different carrier densities.

Figure 7.10a shows the dependence of $h\langle\nu_F\rangle$ on carrier concentration for three different temperatures. Note that the PL stays very close to $kT$ from the band edge, and only increases slightly at higher densities, when band filling takes place.

At the same time, the radiative time drops as the electron concentration increases roughly as $\tau_{rad}(N_e, T) \sim N_e^{-1}$ until the band filling starts to play a significant role (Figure 7.10b). The advent of this effect means that the radiative recombination is no longer bimolecular, as discussed at length above and in [8, 9]. Clearly, the effect of bandtail states on the recombination lifetime is not very strong. Of course, this can easily be understood by simply noting that at low temperatures, the density of "virtual states" is very low in comparison to the density of real states in the bands, while at higher temperatures the density of virtual bandtail states increases, but so does the population of states in the bands. Thus, direct luminescence from the bands always dominates, and the bandtail states only influence the absorption spectra. However, this influence is quite significant, because it affects the choice of the optimum excitation wavelength and thus ultimately the cooling efficiency.

Figure 7.11a plots the value of the shortest break-even recombination time $\tau_{nr,b}$ for a given temperature and concentration. The break-even recombination time clearly first decreases with the electron concentration as the radiative decay rate $\tau_{rad}^{-1}$ increases but then it sharply turns up, although for different reasons. At higher temperatures the increase is caused by the increased Auger recombination, while at lower temperatures the cause is the saturation of absorption and the resulting growth in importance of the background absorption.

In Figure 7.11b the optimal photon energy of the pump laser $h\nu_L$ is plotted. At 100 K the optimum pumping energy is always above the bandgap due to the weakness of the phonon-induced bandtail. At higher temperatures it becomes advantageous to pump into the bandtail states; in other words to use phonon-assisted transitions (as in [23]) for pumping. As the carrier concentration increases the lower-lying phonon-assisted transitions become blocked and it becomes necessary to move the laser energy up and into the band.

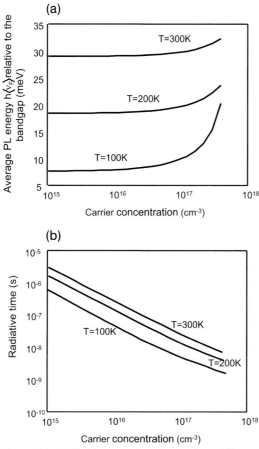

**Figure 7.10** (a) Shift in mean PL energy at three different temperatures. (b) Changes in the radiative lifetime.

The cooling efficiency under the assumption of a relatively long yet attainable nonradiative recombination time $\tau_{nr} = 10\,\mu s$ is shown in Figure 7.12a. As expected, the cooling efficiency peaks at a carrier concentration that is large enough for the radiative recombination to become dominant yet too small to saturate bandtail absorption and make the background absorption a significant factor. These conclusions are no different from those drawn in [8]. The most interesting feature of Figure 7.12, and the one that can be directly attributed to phonon-assisted processes is the nonlinear growth of the maximum cooling efficiency with temperature. Indeed, the maximum cooling efficiency at 200 K, about 1%, is three times higher than at 100 K, while the difference between the efficiencies at 300 K is only a factor of 1.5 – more or less exactly the ratio of temperatures. One can attribute this to the fact that at higher temperatures the optimum pump photon energy lies significantly below the bandgap energy (Figure 7.11b) and thus the quantum efficiency ($\langle \nu_F \rangle - \nu_L)/\nu_L$ experiences a large increase between 100 and 200 K as the pump energy shifts by about 20 meV, but for higher temperatures the $\hbar\nu_L$ stays more or less the same.

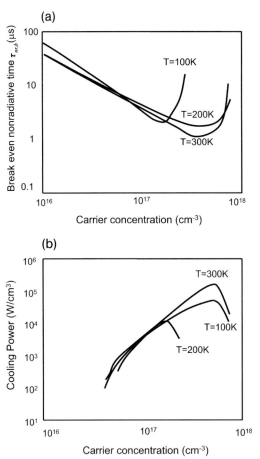

**Figure 7.11** (a) Break-even nonradiative decay time. (b) Optimal photon energy of pump vs. carrier concentration.

In Figure 7.12b the cooling power per unit volume is plotted. As expected, the cooling power increases roughly as the square of the carrier density, and then drops off sharply as the absorption saturates.

These results indicate that the use of phonon-assisted transitions below the gap is conceptually similar to the use of donor–acceptor transitions below the gap, and it offers a similar degree of enhancement in cooling efficiency.

## 7.4
### Laser Cooling Using Type II Quantum Wells

The preceding examples of the engineering of the density of electron–hole pairs in a semiconductor show that improvements in the cooling efficiency can indeed be obtained, but rather unfortunately this improvement is impeded by the fact that the

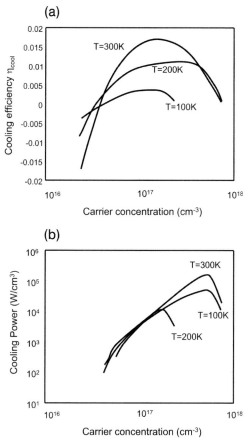

**Figure 7.12** (a) Cooling efficiency. (b) Cooling power density vs. carrier concentration.

bimolecular character of recombination quickly saturates in these materials with a small density of states near laser energy. The saturation of the radiative recombination rate is accompanied by the reduction in absorption at the wavelength of the laser. Therefore, it is desirable to find a material in which the density of states at laser wavelength remains large but the probability of recombination is greatly reduced.

One can consider type II multiple QWs [24], which, in our opinion, satisfies the demands for suppressed photoluminescence at lower energies. This structure, shown in Figure 7.13, consists of alternating layers of lattice-matched InP and $In_{0.53}Ga_{0.47(1-x)}Al_{0.47x}As$. For large Al concentrations ($x > 0.5$), the band alignment favors electron localization and hole localization in different layers. Electrons localize in InP, while holes localize in $In_{0.53}Ga_{0.47(1-x)}Al_{0.47x}As$. Now, the lowest effective bandgap $E_{gap,0}$ is actually an indirect one in real space; this is followed by two direct bandgaps in real space, $E_{gap,1} = E_{gap,0} + \Delta E_v$ and $E_{gap,2} = E_{gap,0} + \Delta E_c$.

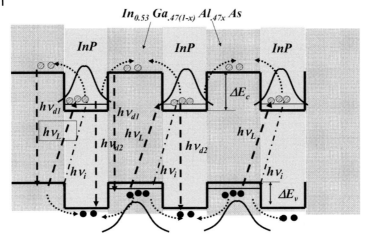

**Figure 7.13** Band diagram of the type II QW structure for efficient laser cooling.

The matrix element of the indirect transition can be evaluated as

$$M_{\text{ind}} = M_{\text{dir}} \int \psi_c(z)\psi_v(z)\,dz, \tag{7.10}$$

where $\psi_c$ and $\psi_v$ are the envelope wavefunctions of the conduction and valence bands, respectively [24]. Therefore, one can write for the radiative lifetime of the spatially indirect transition $\tau_{\text{ind}}^{-1} = \tau_{\text{dir}}^{-1} \times m$, where

$$0 < m = \left| \int \psi_c(z)\psi_v(z)\,dz \right|^2 < 1 \tag{7.11}$$

can be easily varied by changing the period of the MQW structure.

Therefore, one can arrive at a situation with such a small $m$ that most of the radiative recombination occurs via spatially direct transitions at frequencies $\nu_{d,1}$ and $\nu_{d,2}$, and the normalized quantum efficiency of the cooling will be roughly

$$\eta_Q' = \frac{\langle \nu_F \rangle - \nu_L}{kT} \approx 1 + \frac{\frac{\Delta E_v}{kT} e^{-\frac{\Delta E_v}{kT}} + \frac{\Delta E_c}{kT} e^{-\frac{\Delta E_c}{kT}}}{m + e^{-\frac{\Delta E_v}{kT}} + e^{-\frac{\Delta E_c}{kT}}}, \tag{7.12}$$

while the effective radiative lifetime will be

$$\tau_{\text{rad}} \approx \frac{\tau_{\text{rad}}^0}{m + e^{-\frac{\Delta E_v}{kT}} + e^{-\frac{\Delta E_c}{kT}}}. \tag{7.13}$$

In practical terms, since the conduction and valence band offsets are typically quite different, only one of the direct processes is strong. It can then be shown that the quantum efficiency of the process is a solution of a transcendental equation

$$(\eta_Q' - 1) m e^{\eta_Q'} = 1. \tag{7.14}$$

Clearly the maximum improvement of the quantum efficiency of cooling is on the order of $\eta_q \sim -\ln m$ (i.e. a fewfold). This conclusion is confirmed by the exact simulation results shown below.

In Figure 7.14, fluorescence spectra for three different values of InGaAlAs composition are shown at $T = 300$ K for $m = 0.01$. It is apparent that, as Al concentration increases, the direct fluorescence peak shifts towards higher energies. At the same time, the absolute value of the direct peak decreases until, at $x = 0.8$, it becomes comparable to the indirect peak at the band edge.

The quantum efficiency of cooling (7.11) is shown in Figure 7.15 for two different temperatures. For each temperature and relative oscillator strength for the indirect transition there is a QW composition that leads to optimum enhancement of the quantum efficiency of cooling, and the maximum enhancement is indeed on the order of $-\ln(m)$, in accordance with the approximate result of (7.13). This is indeed a very good result, although one should also consider that the total radiative time increases too.

Figure 7.16 shows that the overall radiative recombination rates are indeed greatly reduced in the type II QWs. The critical question for the proposed scheme is thus: what happens to the nonradiative decay in the type II QW? If one considers SRH recombination, one can write

$$\tau_{non}^{-1} \sim \int |\psi_c(z)|^2 |\psi_v(z)|^2 F_t(z)\, dz, \qquad (7.15)$$

where $F_t(z)$ is the density of traps or recombination centers. Assuming that the recombination centers are uniformly spread, we expect a significant reduction in the nonradiative recombination rate. If, on the other hand, the recombination is associated with the interfaces, nonradiative recombination will not be significantly reduced. Since the radiative recombination is reduced, the overall efficiency of cooling will be lowered. Nevertheless, even if we assume no reduction in the nonradiative recombination, we can still achieve a good enhancement of the cooling efficiency, as shown in Figure 7.17.

If we compare the results for type II QWs shown in Figure 7.17 with the ones that can be achieved using the donor–acceptor transitions shown in Figures 7.4

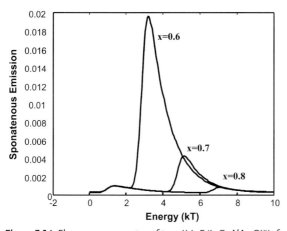

**Figure 7.14** Fluorescence spectra of type II InP/InGaAlAs QWs for different InAlAs fractions.

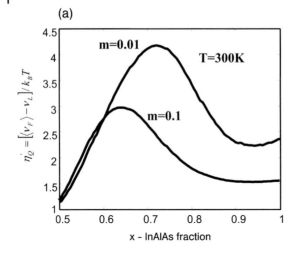
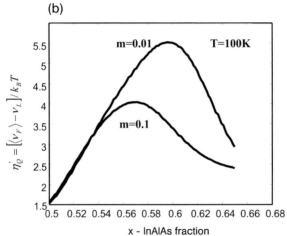

**Figure 7.15** Quantum efficiency of cooling as a function of InAlAs fraction for different values of indirect transition oscillator strength. (a) 300 K, (b) 100 K.

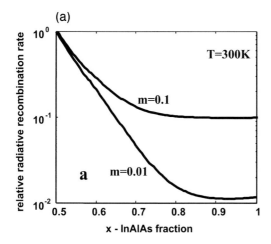

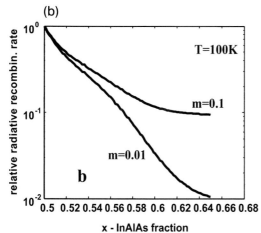

**Figure 7.16** Relative radiative recombination rates as a function of InAlAs fraction for different values of indirect transition oscillator strength. (a) 300 K, (b) 100 K.

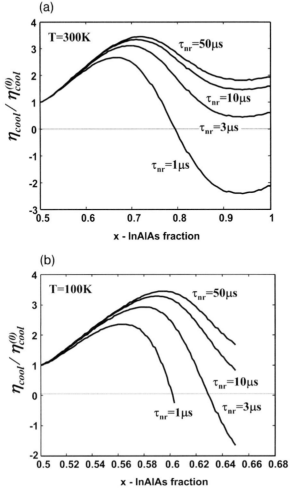

**Figure 7.17** Enhancement in cooling efficiency as a function of InAlAs fraction for different values of indirect transition oscillator strength. (a) 300 K, (b) 100 K.

and 7.5, we can see that the type II results are better – cooling enhancement is possible even with a nonradiative lifetime as short as 1 μs. When this advantage is combined with the relative ease of fabricating type II structures, it becomes clear that, among all of the ways of enhancing cooling efficiency, type II QW structures are probably the way to go.

## 7.5
### Photonic Bandgap for Laser Cooling

The original proposals for photonic crystals in [25, 26] suggested that the photonic bandgap – the range of frequencies in which no electromagnetic wave can prop-

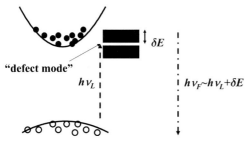

**Figure 7.18** Principle of the use of a photonic bandgap structure for laser cooling.

agate – could be used to inhibit spontaneous radiation in this region. The idea of using the photonic bandgap to facilitate laser cooling naturally follows from this, as can be seen in Figure 7.18. By engineering a photonic bandgap in the vicinity of band-edge transitions of the semiconductor, emission at lower energies becomes forbidden and must take place at higher energies (of the order of $h\nu_F \sim h\nu_L + \delta E$), leading to higher quantum efficiencies. To provide absorption of laser light it is also

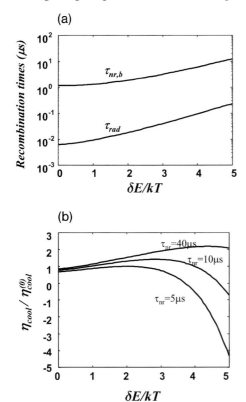

**Figure 7.19** Dependence of various cooling characteristics on the position of the photonic band edge for GaAs at 100 K. (a) Radiative recombination time and critical nonradiative recombination.

necessary to create a "defect band" (i.e. a "waveguide" [27, 28]) in which the pump light can propagate. Reference [26] provides sets of detailed guidelines for designing a photonic crystal with a bandgap of a given width δE and a narrow allowed defect band within it, so we will not concern ourselves with the details of photonic bandgap structure design here. Instead, we will investigate the general features of the use of photonic crystals in laser refrigeration.

The main challenge when using a photonic crystal is to avoid a situation in which the overall radiative recombination rate decreases to the point where nonradiative recombination becomes a significant factor. We have performed detailed modeling and optimization where we varied the position of the upper edge of photonic bandgap relative to the band-edge transition energy in GaAs. The results are shown in Figures 7.19 and 7.20 for 100 and 300 K, respectively.

As one can see from Figures 7.19a and 7.20a, the radiative recombination time increases with the blueshift of the photonic bandgap, causing an increase in break-even nonradiative recombination time. As expected, the increase is somewhat less severe at elevated temperatures, since there are more hot carriers capable of re-

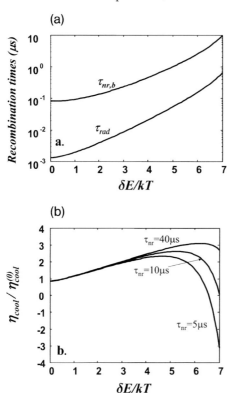

**Figure 7.20** Dependence of various cooling characteristics on the position of the photonic band edge for GaAs at 300 K. (a) Radiative recombination time and critical nonradiative recombination time for $\eta_E = 50\%$. (b) Enhancement of the cooling efficiency for various nonradiative recombination times.

combining at frequencies that are above the photonic bandgap barrier. Plotting the enhancement of the overall cooling efficiency in Figures 7.19b and 7.20b, we can see that efficiency can still be increased by a factor of 2–3 provided that relatively long nonradiative recombination times can be attained.

## 7.6
### Novel Means of Laser Cooling Using Surface Plasmon Polaritons

In this section, we consider an alternative way of simultaneously enhancing and blueshifting the radiative recombination using surface plasmon polaritons (SPPs) [29].

This method was first suggested in [30]. SPPs are the TM waves that propagate along the metal–dielectric interface. All of the interesting SPP phenomena occur in the vicinity of surface plasmon resonance. This happens at a frequency $\nu_{SP}$ where the dielectric constants of the metal and dielectric are equal in magnitude but opposite in sign, $\varepsilon_m(\nu_{SP}) = -\varepsilon_d(\nu_{SP})$. As the frequency asymptotically approaches $\nu_{SP}$, the electric field becomes more and more confined at the interface, which can be expressed as a reduction in the effective width, $w_{eff}$. Concurrently, the propagation constant $\beta$ and its derivative $d\beta/d\omega$ (the inverse of the group velocity) increase. The density of SPP states increases sharply near the SPP resonance $E_{SPP}$, and this leads to the Purcell enhancement of spontaneous emission into SPPs, characterized by the Purcell factor [31]

$$F_p = \frac{\pi\beta}{2w_{eff}} \frac{d\beta}{d\omega} \frac{c^2}{\omega^2 n^3}. \tag{7.16}$$

The idea of using SPPs to enhance the efficiency of spontaneous emission is well known. The density of states in SPP can be quite high, mostly due to its low group velocity; thus Purcell factors in excess of $10^3$ are not unrealistic [32, 33]. The first definite signs of enhancement were attained in GaN photoluminescence upon the addition of a thin Ag film [34]. A 90-fold enhancement of the spontaneous recombination rate in a similar structure was later demonstrated in [32].

Regrettably, the use of SPPs does not necessarily circumvent the "bottleneck" presented by the low density of radiation modes; it simply shifts the bottleneck to the next stage, where the energy needs to be coupled from the high-density SPP modes into the low-density radiation modes. This coupling into the radiation modes must compete with the nonradiative decay of the SPP itself, and this decay occurs at an extremely fast rate. Although various ingenious schemes for coupling SPPs using gratings have been demonstrated [35–37], the overall radiation efficiency of SPP is still low. Hence SPPs can be used to enhance the efficiency of a very weak process, especially a nonlinear one, such as Raman or harmonic generation, but no improvements are expected for semiconductor fluorescence, where the efficiency is already high.

However, the ultimate goal of laser cooling is not the emission of photons externally; it is the transfer of energy from material excitations to the heat sink. Therefore, if an SPP emitted by a recombining electron–hole pair does not transfer its energy into a radiative photon but rather into phonons confined to a metal that is thermally isolated from the semiconductor, the goal of cooling is accomplished. Here we suggest an SPP-assisted laser refrigerating scheme and evaluate the potential improvements in cooling efficiency that can be achieved by using it.

A sketch of the proposed scheme is shown in Figure 7.21. It consists of a thin layer of a semiconductor, GaN in this example, and a heat sink covered with a layer of an SPP-supporting metal, Ag in this example. Also shown is a thin (about 1 nm) layer of a wider bandgap material (AlN in this example), which is introduced to reduce the surface recombination rate. The semiconductor and heat sink are brought close together, but there is still a nanometer-scale vacuum gap of thickness $t_{gap}$ between them. This gap can be maintained using an array of nanometer pillars etched on either the metal or the semiconductor. If the area represented by the pillars is small, the heat sink can be considered to be thermally isolated from the semiconductor. When $t_{gap}$ is much smaller than the wavelength in vacuum, an SPP mode that spreads into the semiconductor can be supported. The field of an SPP mode is also shown in Figure 7.21, while the SPP dispersion curve is plotted in Figure 7.22a for $t_{gap}$ = 10 nm. The dispersion curve shows a significant deviation from the plane wave dispersion in the dielectric (dashed line) in the vicinity of the energy gap of GaN (3.47 eV). However, due to the relatively large imaginary part of the Ag dielectric constant ($\varepsilon_i$ = 0.09 at 3.4 eV), the behavior of the dispersion curve is far from that of the ideal case. Rather than asymptotically approaching a horizontal line, the dispersion turns around at a relatively modest value of $\beta$ equal to just twice the propagation constant of the plane wave. This limits the maximum reduction in the group velocity to a factor of about 20. Furthermore, as $\beta$ increases, the electric field becomes progressively more confined to the vacuum gap, with negative consequences for the Purcell enhancement. Nevertheless, even with all of these limitations, the Purcell factor plotted in Figure 7.22b reaches a substantial value of $F_P \sim 70$ near 3.52 eV before rapidly decreasing. There are two reasons for this sharp drop-off: first, the electric field of the SPP mode becomes confined to the vacuum gap; second, as mentioned above, the group velocity rises rapidly as the dispersion curve in Figure 7.3a turns around.

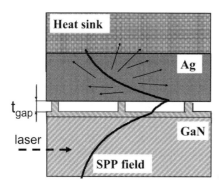

**Figure 7.21** Diagram illustrating SPP-assisted cooling.

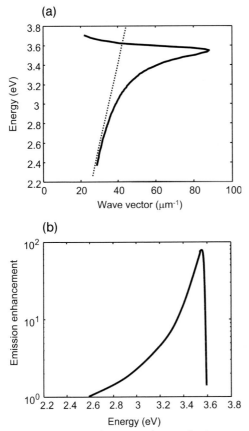

**Figure 7.22** (a) SPP dispersion. (b) Purcell enhancement factor for the structure shown in Figure 7.21.

Next we analyze how one can exploit this Purcell enhancement. In order to attain a desirable blueshift of the photoluminescence, the semiconductor bandgap energy should be positioned a few $k_B T$ below the peak of the Purcell factor. Conveniently, at 300 K, the GaN bandgap energy is 3.4 eV (i.e. about 4 $k_B T$ below the peak), while the GaN bandgap at 100 K occurs at 3.46 eV (i.e. about 6 $k_B T$ below the $F_p$ peak). In principle, the bandgap can be varied by introducing quantum well structures with InGaN wells and AlGaN barriers. Also, the SPP resonance can be shifted to the red part of the spectrum by introducing some kind of profile onto the Ag surface. Reducing the gap thickness also shifts the resonance towards lower energies, but, of course, this reduction requires improvements in fabrication. For now we shall restrict ourselves to the pure binary material – GaN.

To quantify the improvement in cooling attained with SPPs, we have chosen to plot the efficiency of laser cooling, $\eta_{cool}$, against the nonradiative recombination time for two different temperatures, assuming (rather optimistically) that the extraction efficiency without SPPs is 50%. Of course, the extraction efficiency is nearly 100% for SPPs, since practically all SPP scattering takes place in the metal (i.e.

outside the cooled medium). The radiative lifetime in the semiconductor does, of course, depend on carrier concentration. In [10] it was shown that for semiconductors with a relatively narrow gap, such as GaAs, the maximum electron concentration is limited by Auger recombination. In wide-gap semiconductors, however, Auger recombination is not a factor, and the maximum carrier concentration for the purpose of cooling is limited by the absorption saturation that occurs when the difference between the quasi-Fermi levels of the bands approaches the bandgap energy. This restricts the maximum carrier concentration to roughly $10^{17}$ cm$^{-3}$ at 100 K and $6 \times 10^{17}$ cm$^{-3}$ at 300 K.

The results are plotted in Figure 7.23a. In the absence of SPP enhancement (dashed curves), net cooling commences once the nonradiative lifetime exceeds some minimum $\tau_{nonr}^{(min)}$. This threshold time ranges from 90 ns at 100 K to 45 ns at room temperature. As the nonradiative time increases and reaches a few microsec-

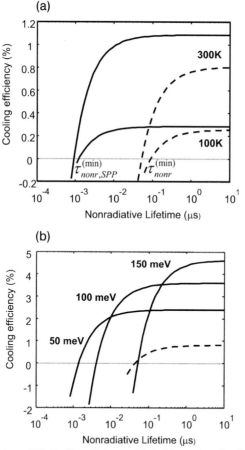

**Figure 7.23** Cooling efficiency with (*solid curves*) and without (*dashed curves*) SPPs vs. nonradiative lifetime: (a) at two different temperatures; (b) at 300 K in the presence of an SPP bandgap for different bandgap energies $E_{g,SPP}$.

onds, the cooling efficiency saturates at roughly $\eta_{\text{cool}}^{(\text{max})} = k_B T/h\nu_L$. Once the SPP scheme of Figure 7.1 is implemented, the efficiency curves shift upward and to the left. The leftward shift is more prominent, indicating that the net cooling can now be obtained even with nonradiative recombination times of less than a nanosecond. This dramatic improvement comes from the fact that SPP emission dominates over nonradiative decay. Recent measurements [38] of various GaN bulk and epitaxial samples have yielded nonradiative times ranging from about 1 ns at room temperature to 10 ns at 100 K. That means that with SPPs, it is possible to attain net cooling even with the (less than perfect) materials that are currently available, while achieving this without SPP would require an improvement of at least two orders of magnitude in nonradiative decay time. There is also an upward shift indicating a modest improvement in maximum cooling efficiency of a few percent at 77 K and 33% at 300 K. This improvement is due to the blueshift of the SPP emission. The reason that this blueshift is rather small can be seen in Figure 7.22b, where the Purcell factor $F_P$ increases by less than an order of magnitude over a range of a few $k_B T$. This broad shape of $F_P(\omega)$ is inherent to lossy material, but can be improved somewhat if the vacuum gap thickness is reduced.

One can also consider combining SPP enhancement with some of the methods of blueshifting considered above (e.g. using type II superlattices). However, the use of SPPs yields new ways of suppressing emission at lower energies. A two-dimensional surface grating – an SPP bandgap structure – can also be introduced [39, 40], which will be able to block SPP emission at energies below the SPP band-edge energy $E_{g,\text{SPP}}$ and thus blueshift the SPP emission spectrum. Such structures have been successfully fabricated by placing periodic triangular arrays of gold bumps on gold surfaces [41]. A truly dramatic increase in efficiency can then be attained, as shown in Figure 7.23b, where we have plotted solid efficiency curves for GaN with an SPP bandgap (three curves are shown, representing three different values of the SPP band-edge energy relative to the GaN band gap energy). Even when the nonradiative lifetime is on the order of 10 ns, the cooling efficiency can exceed 2%.

SPP bandgap structure is far more suitable for laser refrigeration than photonic bandgap structure because it simultaneously suppresses emission at lower energies while enhancing it at higher energies, thus blueshifting the emission peak while keeping the overall emission efficiency high. At the same time, perhaps it can also be said that the reverse is true – that laser cooling is the ideal application for SPPs, because the least attractive feature of SPPs (strong nonradiative energy dissipation) does not factor into laser cooling.

## 7.7
### Conclusions

In this chapter we have considered various means of tailoring the luminescence spectra of semiconductors to facilitate laser cooling. We have shown that to facilitate cooling, it is necessary to move the mean florescence energy as far as possible from the laser energy without adversely affecting the overall recombination rate and while also maintaining sufficient absorption at the wavelength of the laser. We have shown that once all of the factors are taken into account, it is possible to achieve a reasonable trade-off in which the break-even nonradiative rate is decreased and the cooling efficiency is increased a fewfold.

## References

1 Epstein, R.I. et al. (1995) *Nature*, 377, 500.
2 Hoyt, C.W., Sheik-Bahae, M., Epstein, R.I., Edwards, B.C. and Anderson, J.E. (2000) *Phys. Rev. Lett.*, 85, 3600.
3 Thiede, J., Distel, J., Greenfield, S.R. and Epstein, R.I. (2005) *Appl. Phys. Lett.* 86, 154107.
4 Fernandez, J., Garcia-Adeva, A.J. and Balda, R. (2006) *Phys. Rev. Lett.* 97, 033001.
5 Sheik-Bahae, M. and Epstein, R.I. (2004) *Phys. Rev. Lett.*, 92, 247403.
6 Huangm, D., Apostolova, T., Alsing, P.M. and Cardimona, D.A. (2005) *Phys. Rev. B*, 72, 195308.
7 Rupper, G., Kwong, N.H. and Binder, R. (2006) *Phys. Rev. Lett.* 97, 117401.
8 Rupper, G., Kwong, N.H. and Binder, R. (2007) *Phys. Rev. B* 76, 245203.
9 Klingshirn, C.F. (1997) *Semiconductor Optics*, Springer, Berlin, pp 246–248.
10 Khurgin, J.B. (2006) *J. Appl. Phys.*, 100, 113116.
11 Imamoglu, B., Hasselbeck, M.P., Sheik-Bahae, M., Epstein, R.I. and Kurtz, S. (2005) *Appl. Phys. Lett.* 86, 081104.
12 Liebler, J.G., Schmitt-Rink, S. and Haug, H. (1985) *J. Lumin.*, 34, 1.
13 Liebler, J. and Haug, H. (1991) *Europhys. Lett.*, 14, 71.
14 Halperin, B. and Lax, M. (1966) *Phys. Rev.*, 148, 722.
15 Redfield, D. (1963) *Phys. Rev.*, 130, 916.
16 Dexter, D.L. (1967) *Phys. Rev. Lett.*, 19, 383.
17 Dow, J.D. and Redfield, D. (1971) *Phys. Rev. Lett.*, 26, 762.
18 Dunn, D. (1968) *Phys. Rev.* 166, 822.
19 Mahan, G.D. (1966) *Phys. Rev.*, 145, 602.
20 Dow, J.D. and Redfield, D. (1972) *Phys. Rev. B*, 5, 594.
21 Imangholi, B., Hasselbeck, M.P., Sheik-Bahae, M., Epstein, R.I. and Kurtz, S. (2005) *Appl. Phys. Lett.* 86, 081104.
22 Schaefer, W. and Wegener, M. (2002) *Semiconductor Optics and Transport Phenomena*, Springer, Berlin, Chap. 11.
23 Khurgin, J.B. (2008) *Phys. Rev. B*, 77, 235206.
24 Harrison, P. (2005) *Quantum Wells, Wires And Dots*, John Wiley & Sons, Inc., New York.
25 Yablonovitch, E. (1987) *Phys. Rev. Lett.*, 58, 2059.
26 John, S. (1987) *Phys. Rev. Lett.*, 58, 2486.
27 Joannopoulos, J.D., Meade, R.D. and Winn, J.N. (1995) *Photonic Crystals: Molding the Flow of Light*, Princeton Univ. Press, Princeton, pp. 69–72.
28 Mogilevtsev, V., Birks, T.A. and Russel, P.St.J. (1999) *J. Lightwave Tech.*, 17, 2078.
29 Agranovich, V.M., Mills, D.L. (1982) *Surface Polaritons*, North Holland, Amsterdam.
30 Khurgin, J.B. (2007) *Phys. Rev. Lett.*, 98, 177401.
31 Purcell, E.M. (1946) *Phys. Rev.*, 69, 681.
32 Neogi, A. and Lee C.W. et al. (2002) *Phys Rev. B*, 66, 153305.
33 Paella, R. (2005) *Appl. Phys. Lett.*, 87, 111104.
34 Contijo, I., Boroditsky, M. and Yablonovitch, E. et al. (1999) *Phys. Rev. B* 60, 11564.
35 Worthing, P.T. and Barnes, W.L. (2001) *Appl. Phys. Lett.*, 79, 3035.
36 Vuckovic, J., Loncar, M. and Scherer, A. (2000) *IEEE J. Quantum Electron.*, QE-36, 1131.
37 Barnes, W.L. (1999) *IEEE J. Lightwave Tech.*, 17, 2170.

38 CHICHIBU, S.F., UEDONO, A., ONUMA, T. AND NAKAMURA, S. (2005) *Appl. Phys. Lett.* **86**, 021914.

39 KITSON, S.C., BARNES, W.L. AND SAMBLES, J.R. (1996) *Phys. Rev. Lett.*, **77**, 2670–2673.

40 BOZHEVOLNYI, S.I., ERLAND, J. et al. (2001) *Phys. Rev. Lett.*, **86**, 3008–3011.

41 RADKO, I.P., SØNDERGAARD, T. AND BOZHEVOLNYI, S.I. (2006) *Opt. Express*, **14**, 4107–4114.

# 8
# Thermodynamics of Optical Cooling of Bulk Matter
*Carl E. Mungan*

## 8.1
## Introduction

It initially seems surprising that one can optically cool bulk material, be it a condensed sample or a gas of more than the comparatively low number of atoms used in Doppler cooling experiments [1]. Part of the surprise arises from the novelty and is dispelled when one understands that the thermal energy withdrawn from the material is carried away by the radiation emitted by it (and presumably absorbed at some external heat sink that is not in thermal contact with the sample). However, this explanation in terms of the first law of thermodynamics (that is, in terms of the balance between the cooling rate and the net optical power output from the refrigerating sample) is not fully satisfying. Further thought still leaves one perplexed, as it appears that "heat" is being converted into light, the spectrum of which is clearly narrower than that of Planckian thermal radiation, suggesting that entropy is being reduced – violating the second law of thermodynamics. What is missing from the analysis is an accounting of the entropy of the pump source. The reason that one uses a laser to pump the refrigerator (in the case of a photoluminescent cooler) or a current source (for an electroluminescent cooler) is that it is a low-entropy input of energy; ideally a laser beam or electric current is analogous to the "work" used to drive a refrigerator. One can therefore summarize the input and output sources of energy to and from an optical refrigerator by the schematic diagram in Figure 8.1. The principal goal of the present chapter is to quantify these energy and entropy fluxes in order to characterize the refrigeration potential.

The overall organization is as follows. By way of background, Section 8.2 presents a selected review of the history and literature of the thermodynamics of fluorescent cooling of bulk matter. Next, Section 8.3 describes how one relates the entropy to the energy carried by an optical beam and defines various radiation temperatures; specific examples are included to make the formulae concrete. Then Section 8.4 uses those results to calculate the Carnot coefficients of performance of typical solid-state coolers for which actual operating efficiencies have been measured experimentally; various corrections for real-world inefficiencies are also quantified. In

*Optical Refrigeration. Science and Applications of Laser Cooling of Solids.*
Edited by Richard Epstein and Mansoor Sheik-Bahae
Copyright © 2009 WILEY-VCH Verlag GmbH & Co. KGaA, Weinheim
ISBN: 978-3-527-40876-4

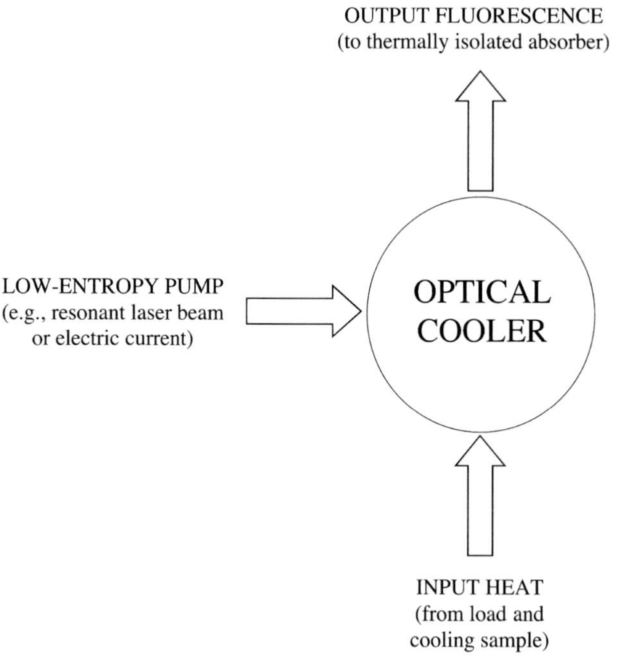

**Figure 8.1** Simplified characterization of the energy input to and output from an optical cooler. Thermal energy is withdrawn from the cooling medium itself, as well as from any external load attached to it. The refrigerator is driven by a low-entropy source, such as a laser resonant with the low-energy wing of an absorption band of the cooling material. Finally, the medium relaxes radiatively and the output fluorescence carries energy away to some external heat sink.

Section 8.5, some key ideas are summarized and the thermodynamics of a few topics related to optical refrigeration are briefly discussed, notably radiation-balanced lasing and the recycling of output optical energy back to the input.

## 8.2
### Historical Review of Optical Cooling Thermodynamics

In a German paper written over three quarters of a century ago, Peter Pringsheim [2] argued that net cooling of a sodium vapor by resonant anti-Stokes emission would not violate the second law of thermodynamics, in contrast to a blanket assertion to the contrary made by Lenard, Schmidt, and Tomaschek. Pringsheim proposed that Na vapor in a glass cell will emit on both yellow $D$ lines when only the lower frequency $D_1$ transition is pumped, as sketched in Figure 8.2. Define the fluorescence quantum efficiency $\eta$ to be the ratio of the number of emitted to absorbed photons (averaged over a long interval compared to the relaxation time $\tau$) and assume it is equal to unity for a sodium vapor at low enough pressure that collisional de-excitations from the upper $P$ to the lower $S$ levels can be neglected.

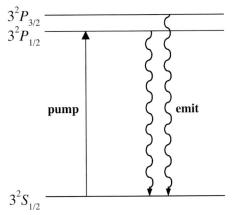

**Figure 8.2** Relevant energy levels of a gas of neutral sodium atoms. First the $D_1$ transition is pumped at 589.6 nm using a spectrally filtered sodium lamp. Then the two upper $P$ levels thermalize with each other, so that they end up with nearly equal population densities (according to the Boltzmann distribution at room temperature). Finally, one gets emission on both the $D_1$ and $D_2$ lines, with the latter having a slightly shorter photon wavelength of 589.0 nm and thus a slightly larger photon energy.

Noting that the average output photon energy is larger than that of the input, one then concludes that the vapor will cool down (until a balance with the heat leak from the surroundings is achieved).

Pringsheim suggested that this optical cooling process can be reconciled with the second law by noting that the sodium vapor is not a closed system – energy is being input in the form of pump light to drive the cooling cycle. But sixteen years later, Vavilov [3] at the Academy of Sciences in Moscow objected that the system is also outputting light, and more radiative energy is leaving it than is entering. Suppose, he argued, that we were to convert the emitted light into electrical energy (using an optical piston for example) and then use that to drive the pump lamp. In that case, it would appear that one could build a self-contained system (lamp, vapor sample, and photovoltaic converter) that transforms thermal energy into useful work in the form of the small excess electrical energy converted. He further pointed out that available experimental data for the fluorescence spectra of organic dyes indicated that the quantum yield $\varrho$ (defined as the ratio of the average fluorescence frequency $\nu_F$ to the pump frequency $\nu_P$) was always less than unity.

Pringsheim [4] replied by noting that the entropy of the emitted radiation is larger than that of the absorbed radiation because the pump light is monochromatic and unidirectional while the fluorescence is broadband and isotropic. One might colloquially say that the latter light waves are more "disordered" (both spectrally and spatially) than the former. (Nowadays, using laser sources, one might add phase coherence to the list of differences between the pump and fluorescence radiation.) Spontaneous luminescence is an intrinsically irreversible process, and so a reversible cooling cycle whose sole effect is the "conversion of heat into work" is not possible. Pringsheim also argued that anti-Stokes fluorescence from dye solutions due to emission between different vibrational levels of two electronic bands can in principle occur with a quantum yield greater than one. If this is not seen in actual experiments, it could be because the fluorescence quantum efficiency is less than one, and it might even be frequency dependent if molecules in the spectral wings are subject to greater perturbations by the solvent than those near the line center.

On the journal pages immediately following Pringsheim's reply are two papers by Vavilov [5] and Landau [6]. In the former, Vavilov makes two new arguments. First, he asserts that the loss of directionality of the fluorescence cannot be associated with an increase in entropy because one can surround the sample with a set of collimating lenses and plane mirrors to steer every emitted ray in essentially the same direction. While true, this argument is nevertheless irrelevant because the entropy of radiation (introduced quantitatively in Landau's paper) is actually an integral over the product of a beam's cross-sectional area and solid-angle divergence (and not over the latter alone), and – according to Liouville's theorem – that brightness product cannot be decreased by a passive collection of lenses and mirrors [7]. Second, Vavilov noted the conflicting requirements of making the pressure of the sodium gas low enough to obviate nonradiative relaxation from the $P$ down to the $S$ levels while keeping the pressure high enough to ensure thermal equilibration between the two excited $P$ levels. Since both processes are mediated by atomic collisions, it is clear that one cannot simultaneously satisfy these two requirements perfectly. While admitting Vavilov's point, it nonetheless cannot be concluded that fluorescent cooling of a gas is infeasible – as a counterexample, modern experiments have demonstrated the anti-Stokes cooling of carbon dioxide [8, 9]. In retrospect, one can argue that the energy exchanged between two colliding atoms can at most be of order $kT = 25$ meV at room temperature (where $k$ is Boltzmann's constant). This estimate implies that a single collision can readily transfer atoms from either excited $P$ level to the other. In contrast, the energy gap between these excited levels and the ground $S$ state is about $80\,kT$, which means that de-excitation by collisions between sodium atoms is strongly suppressed. (On the other hand, collisional relaxation at the walls of the gas cell is much more likely because of the large effective spring constant of the matrix-bonded glass atoms. In fact, in the carbon dioxide experiments mentioned above, cooling only occurs along the central axis of the long cylindrical cell, well away from its curved surface.) The point is that there is a definite crossover region in the gas density: high enough that the rate of collisions between pairs of atoms is large compared to the radiative relaxation rate $1/\tau_R$ but low enough that simultaneous collisions between many atoms seldom occur.

Interestingly enough, if we assume that the cooling coefficient of performance $\kappa$ is roughly equal to the standard Carnot value for a refrigerator,

$$\kappa_C = \frac{T_L}{T_H - T_L}, \qquad (8.1)$$

where the low-temperature reservoir is taken to be the optical cooler operating near room temperature, $T_L = 300$ K, and the hot reservoir is taken to be the "effective temperature" (whose meaning is clarified below) of the waste fluorescence, with

$T_H$ estimated by Landau [6] to be say 10 000 K, then one finds $\kappa \approx 3\%$. This is remarkably close to the measured values of the best currently known optical coolers, as recently reviewed by Mungan et al. [10]. Therefore despite Vavilov's objections, one can reasonably assert that the basic theoretical validity of the concept of optical cooling of bulk matter was already established in his era. However, practical implementations of the idea, pioneered by the experiments of Kushida and Geusic [11] using $Nd^{3+}$:YAG, had to await the invention of the laser.

A more accurate expression for the Carnot coefficient of performance (COP) than (8.1) follows from the work of Geusic, Schulz-DuBois, and Scovil [12]. They point out that there are actually three temperatures (not just two) to be considered: the temperature of the cooling sample $T$, the effective temperature of the fluorescence $T_F$, and the effective temperature of the pump $T_P$. As above, we assume that $T_F$ is substantially larger than $T$. In addition, since the pump radiation must have much lower entropy flux than the fluorescence (in order to satisfy the second law of thermodynamics) with approximately the same power, it follows that $T_P \gg T_F$. One can therefore schematize the situation with three thermal reservoirs arranged in vertically decreasing order of temperature, as in Figure 8.3. A refrigerator operates between the lower two reservoirs, with a Carnot COP of

$$\kappa_{\text{fridge, C}} = \frac{T}{T_F - T} \qquad (8.2)$$

by substituting the appropriate high and low temperatures into (8.1). The work $W$ required to drive the operation of this fridge comes from a heat engine operating between the upper two reservoirs. The usual expression for its Carnot efficiency is

$$\varepsilon_{\text{engine, C}} = \frac{T_P - T_F}{T_P} . \qquad (8.3)$$

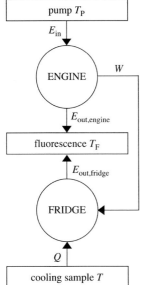

**Figure 8.3** A heat engine and a refrigerator coupled in tandem between three thermal baths as a model for an optical cooler. In one excitation–relaxation cycle, the energy $E_{in}$ is absorbed from the pump source, heat $Q$ is withdrawn from the cooling sample (and its load), and net energy $E_{out} \equiv E_{out,engine} + E_{out,fridge}$ is exhausted in the form of fluorescence. All of the work $W$ output from the engine is used to drive the fridge.

By multiplying together (8.2) and (8.3), we obtain the overall Carnot coefficient of performance of the optical cooler,

$$\kappa_C = \frac{T - \Delta T}{T_F - T}, \tag{8.4}$$

where $\Delta T \equiv T T_F/T_P$ represents a correction to the temperature in the numerator, slightly decreasing the COP compared to what it otherwise would have been. An analogous expression can be deduced for the Carnot efficiency of an optically pumped laser, which is essentially an optical cooler running in reverse [13–15]. Note that $T_P \to \infty$ for an ideal laser or electric-current pump source, in which case (8.4) reduces to

$$\kappa_C = \frac{T}{T_F - T}, \tag{8.5}$$

in agreement with the preceding discussion of (8.1). This Carnot COP falls approximately linearly to zero as $T \to 0$ starting from a sample temperature substantially below $T_F$.

Now consider coupling these temperature baths to a nondegenerate three-level system like that of Figure 8.2, resulting in the arrangement sketched in Figure 8.4. Let the population densities in the three levels be $N_1$, $N_2$, and $N_3$. Since levels 2 and 3 thermalize with the sample, their population ratio follows a Boltzmann distribution,

$$\frac{N_3}{N_2} = e^{-h\nu_{23}/kT}, \tag{8.6}$$

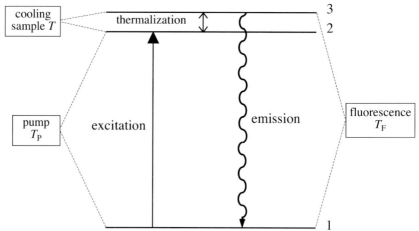

**Figure 8.4** Coupling of a three-level system to a set of thermal reservoirs that maintain Boltzmann population ratios at the appropriate temperatures and transition frequencies. In order to maximize the cooling efficiency, spontaneous emission between levels 2 and 1 has been neglected.

where $h$ is Planck's constant and where the transition frequency between levels 2 and 3 is the difference between the emission and excitation frequencies, $\nu_{23} = \nu_F - \nu_P$. Let us similarly assume that the pump source consists of a spectrally filtered arc discharge lamp that is coupled to the system in such a fashion that we get a Boltzmann population ratio between levels 1 and 2 that is determined by the very high (thousands of degrees) temperature $T_P$ of the arc,

$$\frac{N_2}{N_1} = e^{-h\nu_P/kT_P}. \tag{8.7}$$

Multiplying (8.6) and (8.7) together and setting the result equal to

$$\frac{N_3}{N_1} = e^{-h\nu_F/kT_F} \tag{8.8}$$

defines an effective temperature of the fluorescence, $T_F = TT_P\nu_F/(T_P\nu_F - T_P\nu_P + T\nu_P)$. Substituting this result into (8.4) in the form $\kappa_C = (1 - T_F/T_P)/(T_F/T - 1)$ gives rise to

$$\kappa_C = \frac{\nu_F - \nu_P}{\nu_P} = \frac{\lambda_P - \lambda_F}{\lambda_F}, \tag{8.9}$$

where in the second equality the pump and mean fluorescence vacuum wavelengths are the speed of light $c$ divided by the corresponding frequencies. Note that the first equality accords with the definition of the COP as $Q/E_{in}$, the ratio of the net cooling energy to the (absorbed) pump energy per cycle, so that Figures 8.3 and 8.4 are consistent with each other. (One can also express this equality as $\kappa_C = \varrho - 1$ in terms of the quantum yield discussed above.) This result assumes that the heat engine and refrigerator in Figure 8.3 are ideal devices, so that entropy is conserved. In practice, however, fluorescence is a spontaneous (irreversible) process and the energy transfers to and from the thermal reservoirs do not proceed quasistatically, which means that the actual coefficient of performance cannot attain the Carnot value given by (8.4).

To numerically estimate this ideal cooling efficiency, we can approximate the difference between the emission and excitation photon energies in Figure 8.4 by a thermally absorbed energy of $kT$. In that case, (8.9) becomes $\kappa_C \approx kT\lambda_P/hc$, which at room temperature is equal to 2% for 1-μm excitation. This value is in good agreement with experimental measurements on the laser cooling of ytterbium doped in a heavy-metal-fluoride glass [16].

Landau [6] defined the effective temperature $T_F$ of the fluorescence by what is nowadays called its "brightness temperature", which is the temperature of a black body whose spectral radiance is equal to that of the fluorescence averaged over its bandwidth (assumed to be narrow). This is also the radiation temperature used by Ross [17], but he distinguished it in general from temperature $T_H$ owing to energy losses. Weinstein [18] appears to have been the first person to define the effective temperature instead in terms of a second, distinct quantity that has been dubbed the "flux temperature" by Landsberg and Tonge [19]. The flux temperature is the ratio of the energy and entropy carried by a beam of light. (Careful mathematical definitions of the brightness and flux temperatures appear in Section 8.3 below.) Weinstein used (8.5) to compute the maximum visible emission efficiency, $1 + \kappa_C$, of an electrically pumped lamp or phosphor.

A more logical choice of electroluminescent cooler than a lamp or phosphor is a semiconductor diode. It was noted in the early 1950s that the threshold voltage for recombination emission across a p–n junction in silicon carbide is slightly smaller than its bandgap, which could therefore lead to a cooling effect [20]. In brief follow-up papers, Tauc [21] and Gerthsen and Kauer [22] argued that the idealized cooling COP is

$$\kappa = \frac{E_g - eV}{eV}, \qquad (8.10)$$

where $E_g$ is the bandgap energy, $e$ is the electron charge, and $V$ is the forward bias voltage. This expression assumes unit external fluorescence quantum efficiency and zero Joule heating. If we identify the mean emission frequency as $\nu_F = E_g/h$ and the pump energy as $h\nu_P = eV$ per electron, then (8.10) is seen to be the direct analog of (8.9). However, it is more accurate to obtain $\nu_F$ from an actual measurement of the emission spectrum (rather than assuming it is equal to the bandgap frequency), in which case we can rewrite (8.10) as

$$\kappa = \frac{h\nu_F}{eV} - 1. \qquad (8.11)$$

For example, if one estimates $\nu_F$ to be the near-infrared peak emission frequency of a GaAs diode operating at 78 K [23], one finds that $\kappa$ is 3% for a 1.335-V bias. A detailed thermodynamic analysis of this example was conducted by Nakwaski [24].

## 8.3
### Quantitative Radiation Thermodynamics

The directional density of states (number of modes per unit volume in a frequency interval $d\nu$ and element of solid angle $d\Omega = \sin\theta\, d\theta\, d\phi$) for a beam of photons in vacuum is [6]

$$G_{\nu,\Omega} = 2\frac{\nu^2}{c^3}, \qquad (8.12)$$

where the factor of two arises from the two independent transverse polarizations of light. If the radiation propagates isotropically in all directions, then one can integrate (8.12) over all solid angles to obtain $G_\nu = 8\pi\nu^2/c^3$, which is the familiar formula for the electromagnetic mode density used in the derivation of the Planck distribution [25] and obtained by counting the number of standing waves in a cavity. Consequently, the number of modes per unit time is $cG_{\nu,\Omega}\, dA_\perp\, d\nu\, d\Omega$, where $dA_\perp = \cos\theta\, dA$ is the element of surface area $dA$ projected into the direction of photon propagation (specified by polar and azimuthal angles $\theta$ and $\phi$, respectively, in spherical coordinates) in Figure 8.5. The radiation is distributed over these

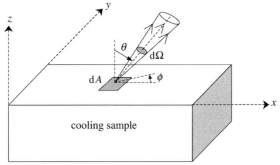

**Figure 8.5** Portion of the radiation emitted from surface area $dA$ of the sample into solid angle $d\Omega$. The polar angle $\theta$ is measured relative to the surface normal direction defining the z-axis, while the azimuthal angle $\phi$ is measured counterclockwise from the x-axis aligned along any convenient direction tangential to the sample surface.

modes with an occupation number $n$ (not to be confused with refractive index) that depends on $\nu$, $\theta$, $\phi$, and the two position coordinates $x$ and $y$ on the surface $A$ (which, for example, can be taken to span the faces of the optical cooling sample as it emits fluorescence). Multiplying the number of occupied modes per unit time by the energy $h\nu$ per photon and integrating gives the optical power,

$$\dot{E} = 2hc^{-2} \int_A \int_\Omega \int_\nu n\nu^3 \, d\nu \cos\theta \, d\Omega \, dA , \tag{8.13}$$

where the overdot denotes a time derivative of the energy $E$ carried by the beam, assumed to be unpolarized and continuous wave. The frequency and angular integrations are respectively over the spectral peaks and range of solid angles (for example, $2\pi$ in Figure 8.5) relevant to the absorption or emission process of interest. A related quantity is the spectral radiance $L_\nu \equiv d\dot{E}/dA_\perp \, d\nu \, d\Omega$, sometimes called the brightness [26],

$$L_\nu = \frac{2nh\nu^3}{c^2} . \tag{8.14}$$

The radiance is $L \equiv \int L_\nu \, d\nu$; in general one defines the derivative of a quantity with respect to frequency or wavelength as the corresponding "spectral" quantity, subscripted with $\nu$ or $\lambda$. The occupancy $n$ determines key parameters of the radiation, including its energy and entropy fluxes and effective temperatures. In turn, $n$ can be obtained experimentally from (8.14) by measuring the spectral radiance.

The brightness temperature $T_b$ of radiation from some source is formally defined as the temperature of a black body such that the spectral radiances of the Planck spectrum and of the source are equal to each other when they are averaged over some narrow range of frequencies $\delta\nu$, solid angle $\delta\Omega$, and area $\delta A$. Since blackbody photons follow the Bose–Einstein distribution [25], (8.14) therefore implies that

$$\int_{\delta A}\int_{\delta\Omega}\int_{\delta v} nv^3\, dv \cos\theta\, d\Omega\, dA = \int_{\delta A}\int_{\delta\Omega}\int_{\delta v} \frac{v^3}{\exp(hv/kT_b)-1}\, dv \cos\theta\, d\Omega\, dA. \tag{8.15}$$

For narrowband radiation with a central frequency $v_0 = c/\lambda_0$, such as might be emitted by an LED or laser whose average spectral radiance is $\bar{L}_v \equiv L/\Delta v$ where $\Delta v$ is the bandwidth of the radiation, (8.14) implies that the average photon occupation number is $\bar{n} \approx c^2 \bar{L}_v/2hv_0^3$. Substituting this result into (8.15) leads to a mean brightness temperature of

$$\bar{T}_b \approx \frac{hv_0}{k\ln(1+1/\bar{n})}, \tag{8.16}$$

which is therefore determined by the peak frequency and mean occupation number. For a very bright source, (8.16) reduces to the particularly simple form $k\bar{T}_b \approx \bar{n}hv_0 \approx \lambda_0^2 \bar{L}_v/2$. If the radiation is emitted from an area $A$ into a circular cone of divergence half-angle $\delta$, then the spectral radiance averaged over area, frequency, and angles is

$$\bar{L}_v = \frac{\dot{E}}{\int_0^{2\pi}\int_0^{\delta}\int_{\Delta v}\int_A \cos\theta\, dA\, dv\, \sin\theta\, d\theta\, d\phi} = \frac{\dot{E}}{A\,\Delta v\,\pi \sin^2\delta}. \tag{8.17}$$

For example, an unpolarized 1 mW red He-Ne laser with a beam area of 1 mm², a divergence of 0.5 mrad corresponding to a solid angle of approximately $\pi\delta^2 = 0.8\,\mu\text{sr}$, and a bandwidth of 1 GHz has a mean brightness temperature of $2\times 10^{10}$ K. Assuming Gaussian spectral and angular profiles with cylindrical symmetry, the detailed variation of $T_b$ with $\theta$ and $v$ has been plotted in Figure 8.2 of Essex et al. [27]. For uniform emission over a hemisphere ($\delta = \pi/2$), note that (8.17) becomes $\dot{E}/A\,\Delta v\,\pi$ and not $\dot{E}/A\,\Delta v\,2\pi$, as one might have naively assumed by dividing the optical power $\dot{E}$ by the emitting surface area $A$, the frequency bandwidth $\Delta v$, and the solid angle $2\pi$ of half of a sphere. This factor of two difference is a result of Lambert's cosine law [28].

Before continuing, it is appropriate to consider in more detail how to characterize a spectral quantity $F_\lambda$ such as radiance or flux, illustrated in Figure 8.6. The dominant wavelength $\lambda_0$ is defined to be the centroid of the spectrum,

$$\lambda_0 = \frac{\int \lambda F_\lambda\, d\lambda}{\int F_\lambda\, d\lambda}. \tag{8.18}$$

Noting that $F_\lambda\, d\lambda = F_v\, dv$, then, in the case of the spectral energy flux density $I_{E\lambda} \equiv d\dot{E}/dA\, d\lambda$ (loosely called the "intensity" when a single-beam scan is recorded by a spectrometer), the mean fluorescence photon frequency is

$$v_F \equiv \frac{\int v\,(I_{Ev}/hv)\, dv}{\int (I_{Ev}/hv)\, dv} \Rightarrow v_F^{-1} = \frac{\int v^{-1} I_{Ev}\, dv}{\int I_{Ev}\, dv}, \tag{8.19}$$

(a)

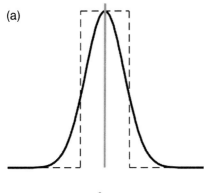

**Figure 8.6** Three examples of plots of some spectral quantity $F_\lambda$ versus wavelength $\lambda$. Each is characterized by three values: $\lambda_0$ denoted by the *solid vertical line* and equal to the centroid of the spectrum; $F_0$, indicated by the *height of the dashed rectangle* and equal to the peak value of $F_\lambda$; and $\Delta\lambda$ represented by the *width of the dashed rectangle* and chosen so that the areas under the spectrum and under the dashed rectangle are equal. (a) A single Gaussian peak. (b) A series of three spikes that might represent emission from a multimode diode laser. (c) Sum of a pair of Lorentzians composing a peak with a long-wavelength shoulder.

(b)

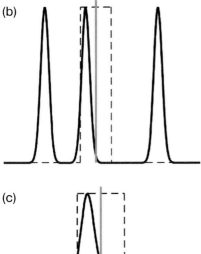

(c)

because $I_{E\nu}/h\nu$ is the photon spectral flux density in SI units of $s^{-1}m^{-2}Hz^{-1}$ [29]. By inspection, the second equality is seen to agree with (8.18) when $F_\lambda = I_{E\lambda}$ and $\lambda_0 \equiv \lambda_F = c/\nu_F$. For a narrowband fluorescence spectrum, however, only a small error is made if one instead writes (8.18) with $\lambda$ replaced by $\nu$, which is equivalent to "canceling" the two factors of $h\nu$ in the first equality or the reciprocals in the second equality of (8.19). Next, if we define $F_0$ to be the maximum value of $F_\lambda$ (after suitable smoothing or averaging), then $\Delta\lambda$ can be defined so that the peak-bandwidth product is

$$F_0 \Delta\lambda = \int F_\lambda \, d\lambda . \qquad (8.20)$$

Let's apply these ideas to the fluorescence emitted by an $Yb^{3+}$:ZBLAN optical cooler pumped with $\dot{E}_P = 40$ W [15]. The fluorescence power is $\dot{E}_F = (1 + \kappa)\dot{E}_P \approx \dot{E}_P$ since the coefficient of performance $\kappa$ that has been experimentally observed is at most 3%. We take the cooling sample to be cylindrical with a length of 3 cm and a radius of 1.5 cm, so that its surface area is $A_{sample} = 13.5\pi$ cm$^2$. The fluorescence spectrum at room temperature is similar in appearance to Figure 8.6c and is experimentally observed to have a mean wavelength of $\lambda_F = 995$ nm and a bandwidth of $\Delta\lambda_F = 35$ nm, computed from (8.20) as the integral of the normalized spectral irradiance (also called the lineshape profile g),

$$\Delta\lambda = \int \frac{I_{E\lambda}}{I_{E0}} d\lambda \equiv \int g(\lambda)\, d\lambda . \tag{8.21}$$

Noting that frequency is related to vacuum wavelength by $\nu = c/\lambda$, it follows that $d\nu = c\, d\lambda/\lambda^2$ in magnitude, and hence for narrowband light that

$$\frac{\Delta\lambda}{\lambda_0} = \frac{\Delta\nu}{\nu_0}, \tag{8.22}$$

from which one concludes that $\Delta\nu_F = c\Delta\lambda_F/\lambda_F^2$. Substituting this result and (8.17) into the average of (8.14) then implies that

$$\bar{n}_F = \frac{\lambda_F^5 \dot{E}_F}{2\pi hc^2 \Delta\lambda_F A_{sample}} \tag{8.23}$$

assuming that the fluorescence is emitted homogeneously and hemispherically from the sample surface, although the detailed distribution of the light is actually nonuniform, as analyzed by Chen et al. [30] using ray tracing. Inserting the numbers given above, one then finds the moderately small value $\bar{n}_F = 7 \times 10^{-4}$, and when this is substituted into (8.16) a fluorescence brightness temperature of 2000 K is calculated. The physical significance of this result can be understood from Figure 8.7, which compares the spectral radiance of a black body (BB),

$$L_\lambda^{BB} = \frac{2hc^2/\lambda^5}{\exp(hc/kT_b\lambda) - 1}, \tag{8.24}$$

at a temperature of $T_b = 2000$ K with that of the fluorescence idealized as a single Gaussian,

$$L_\lambda = L_0 \exp\left[-\frac{1}{2}\left(\frac{\lambda - \lambda_0}{w}\right)^2\right]. \tag{8.25}$$

The peak value $L_0$ and the spectral width $w$ corresponding to one standard deviation can be obtained by rewriting (8.21) in two different ways after substituting $I_{E\lambda}/I_{E0} = L_\lambda/L_0$ into it. On the one hand, we have

$$\Delta\lambda = \frac{\bar{L}_\lambda \Delta\lambda}{L_0} \Rightarrow L_0 = \frac{\dot{E}_F}{\pi \Delta\lambda_F A_{sample}} \tag{8.26}$$

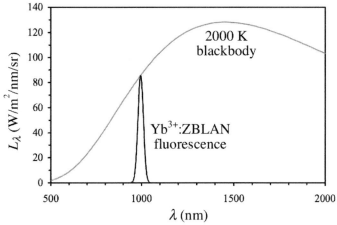

**Figure 8.7** Comparison of the spectral radiance of a 2000 K black body and of the ytterbium-doped heavy-metal-fluoride fluorescence (approximated as a Gaussian peaking at 995 nm with a 35 nm bandwidth) corresponding to a hemispherical emittance of $\dot{E}_F/A_{sample} = 0.94$ W/cm$^2$.

using (8.17) to get the second equality. (The fact that the peak and mean values are equal is simply a consequence of the operational definition of the average of a spectral quantity given above, $\bar{L}_\lambda \equiv \int L_\lambda \, d\lambda / \Delta\lambda$. Theoretically it is preferable to integrate the numerator over the entire spectrum, rather than just over the bandwidth as in [31], because the result can then be related to the oscillator strength.) On the other hand, we can integrate (8.25) over all wavelengths to conclude that

$$\Delta\lambda = w\sqrt{2\pi} \approx 2w\sqrt{2\ln 2}, \tag{8.27}$$

where the last expression defines the full width at half-maximum (FWHM) of the Gaussian, in agreement with an inspection of Figure 8.6a. Therefore $w = \Delta\lambda_F/\sqrt{2\pi}$. As illustrated in Figure 8.7, the upshot is that one can find the brightness temperature of a narrowband source by plotting its spectral radiance and determining the black-body curve that passes through its peak.

In contrast to the brightness temperature, computation of the flux temperature first requires the calculation of the rate $\dot{S}$ at which entropy is carried by the radiation. Start by defining a (macro) state of a system of photons as consisting of $N_1$ photons in optical mode 1, $N_2$ photons in optical mode 2, and so on, not to be confused with the population densities elsewhere in this chapter. (For the moment, we are treating the modes as though they were discrete. At the end of the calculation, we will generalize the result to a continuous distribution.) Here an optical mode is defined by a particular set of values of $\nu$, $\theta$, $\phi$, $x$ and $y$, as discussed prior to (8.13). Now consider some ensemble of a large number $M$ of systems [27]. The entropy of the entire ensemble is $MS$. The probability that we find the system in some particular state is $P_{state} \equiv P(N_1, N_2, \ldots)$. Therefore, the number of systems we will find in that state is $m_{state} = MP_{state}$. If we label the states of the system by $A, B, \ldots$, then the number of ways that the systems can be arranged to form the given ensemble is

$$W = \frac{M!}{m_A! m_B! \ldots} . \tag{8.28}$$

Hence the entropy of this ensemble of systems is

$$MS = k \ln W \approx k \left[ M \ln M - \sum_{\text{states}} m_{\text{state}} \ln m_{\text{state}} \right] \tag{8.29}$$

using Stirling's approximation in the second step. (The sum need only run over *accessible* states. Stirling's approximation for $m_{\text{state}}!$ can then always be made valid by choosing $M$ large enough.) Substituting $m_{\text{state}} = M P_{\text{state}}$ and noting that the summation of $P_{\text{state}}$ over all possible states of the system must be unity, we conclude that

$$S = -k \sum_{\text{states}} P_{\text{state}} \ln P_{\text{state}} = -k \sum_{N_1=0}^{\infty} \sum_{N_2=0}^{\infty} \ldots P(N_1, N_2, \ldots) \ln P(N_1, N_2, \ldots) , \tag{8.30}$$

which is called the Shannon entropy [32] of the system of photons. Next assume that the probability $p_i(N_i)$ of finding $N_i$ photons in mode $i$ is independent of the probability $p_j(N_j)$ of finding $N_j$ photons in mode $j$ if $i \neq j$ [19]. In that case, the probability of finding the system in a particular state is given by the product

$$P(N_1, N_2, \ldots) = \prod_{i=1}^{\infty} p_i(N_i) , \tag{8.31}$$

so that

$$-S/k = \sum_{N_1=0}^{\infty} \sum_{N_2=0}^{\infty} \sum_{N_3=0}^{\infty} \ldots (p_1 p_2 p_3 \ldots)(\ln p_1 + \ln p_2 + \ln p_3 + \cdots)$$

$$= \sum p_1 \ln p_1 \sum p_2 \sum p_3 \cdots + \sum p_1 \sum p_2 \ln p_2 \sum p_3 \cdots$$

$$+ \sum p_1 \sum p_2 \sum p_3 \ln p_3 \cdots + \cdots . \tag{8.32}$$

However, each summation over $p_i$ alone is equal to unity by the normalization condition,

$$\sum_{N_i=0}^{\infty} p_i(N_i) = 1 . \tag{8.33}$$

Therefore (8.32) can be tidily expressed as

$$S = \sum_{i=1}^{\infty} S_i , \tag{8.34}$$

where the partial entropy [31] of mode $i$ is

$$S_i \equiv -k \sum_{N_i=0}^{\infty} p_i(N_i) \ln p_i(N_i) . \tag{8.35}$$

This result states that the entropy of electromagnetic radiation is a sum of the Shannon entropy of each optical mode. To continue, let's further assume that the probability of finding one additional photon in a mode is independent of the number of photons already occupying that particular mode. This assumption implies that

$$p_i(N_i) \propto q_i^{N_i} , \tag{8.36}$$

where $q_i$ is the relative probability of finding one more photon in mode $i$. The normalization of (8.36) according to (8.33) becomes a geometric series, resulting in

$$p_i(N_i) = (1 - q_i) q_i^{N_i} . \tag{8.37}$$

However, the (mean) occupation number of mode $i$ is

$$n_i = \sum_{N_i=0}^{\infty} N_i p_i(N_i) = (1 - q_i) q_i \frac{d}{dq_i} \sum q_i^{N_i} = \frac{q_i}{1 - q_i} . \tag{8.38}$$

For example, $q_i$ is equal to $\exp(-h\nu_i/kT)$ for radiation in thermal equilibrium at temperature $T$ (owing to the equal spacing of the energy levels of the black-body oscillators with which the radiation interacts), so that the (average) number of photons in mode $i$ becomes

$$n_i = \frac{1}{\exp(h\nu_i/kT) - 1} . \tag{8.39}$$

Dropping the subscript $i$, this result is the familiar Planck occupation number [33] used in (8.15) above.

Returning to the general case of nonequilibrium radiation, (8.38) can be inverted to give

$$q_i = \frac{n_i}{1 + n_i} . \tag{8.40}$$

Substituting this result into (8.37), and then that into (8.34) and (8.35), one obtains

$$S = -k \sum_{i=1}^{\infty} \sum_{N_i=0}^{\infty} \frac{n_i^{N_i}}{(1 + n_i)^{N_i+1}} \ln \frac{n_i^{N_i}}{(1 + n_i)^{N_i+1}}$$

$$= k \sum_{i=1}^{\infty} \sum_{N_i=0}^{\infty} \frac{(N_i + 1) n_i^{N_i}}{(1 + n_i)^{N_i+2}} (1 + n_i) \ln(1 + n_i) \tag{8.41}$$

$$- k \sum_{i=1}^{\infty} \sum_{N_i=1}^{\infty} \frac{N_i n_i^{N_i-1}}{(1 + n_i)^{N_i+1}} n_i \ln n_i .$$

Let $N = N_i$ in the first summation and $N = N_i - 1$ in the second one, to get

$$S = k \sum_{i=1}^{\infty} \left[(1+n_i)\ln(1+n_i) - n_i \ln n_i\right] \left\{ \frac{1}{(1+n_i)^2} \frac{d}{dq_i} \sum_{N=0}^{\infty} q_i^{N+1} \right\}, \tag{8.42}$$

where the quantity in the curly brackets equals unity. Therefore, the partial entropy for one mode of occupation number $n$ is $k[(1+n)\ln(1+n) - n \ln n]$. Finally, as in the computation of (8.13), we multiply this result by $cG_{\nu,\Omega}\,dA_\perp\,d\nu\,d\Omega$ and integrate to find

$$\dot{S} = 2kc^{-2} \int_A \int_\Omega \int_\nu \left[(1+n)\ln(1+n) - n \ln n\right] \nu^2 \, d\nu \cos\theta \, d\Omega \, dA, \tag{8.43}$$

which is called the entropy flux.

For narrowband radiation that is independent of the angular directions of propagation $\theta$ and $\phi$ within a circular cone of half-angle $\delta$, then its energy flux density $I_E \equiv d\dot{E}/dA$ is

$$I_E \approx 2\pi h c^{-2} \bar{n} \nu_0^3 \Delta\nu \sin^2\delta, \tag{8.44}$$

from (8.13), while its entropy flux density $I_S \equiv d\dot{S}/dA$ using (8.43) is

$$I_S \approx 2\pi k c^{-2} \left[(1+\bar{n})\ln(1+\bar{n}) - \bar{n}\ln\bar{n}\right] \nu_0^2 \Delta\nu \sin^2\delta. \tag{8.45}$$

Because the thermodynamic definition of temperature is $1/T \equiv \partial S/\partial E$ at constant volume, which we can take to be that of the sample considered as an "optical converter" of radiation from one form into another [19], it follows that

$$T = \frac{dI_E}{dI_S} = \frac{h\nu_0 \, d\bar{n}}{k\ln(1+1/\bar{n})\,d\bar{n}} = \bar{T}_b, \tag{8.46}$$

using (8.16) in the last step. Thus the brightness temperature is an absolute thermodynamic temperature even for nonequilibrium radiation.

If either the bandwidth $\Delta\nu$ or the divergence $\delta$ of the light approaches zero in a manner such that its energy flux density remains finite, then (8.44) requires that the mean occupation number $\bar{n} \to \infty$. In these limits, (8.45) becomes

$$I_S \approx \frac{k}{h\nu_0} I_E \frac{\ln\bar{n}}{\bar{n}} \to 0. \tag{8.47}$$

That is, if the radiation crossing a differential area $dA$ is either monochromatic or unidirectional, as illustrated in Figure 8.8, then the beam carries zero entropy. In this sense, one can characterize an ideal laser beam as purely doing work to drive an optical cooler, in striking contrast to black-body radiation which only delivers heat [15]. Note however from (8.46) that $\int T\,dI_S$ is always equal to the irradiance and does *not* define a heat flux density for nonequilibrium radiation.

The flux temperature of radiation is now defined as

$$T_f \equiv \frac{I_E}{I_S}, \tag{8.48}$$

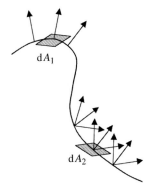

**Figure 8.8** Distinction between unidirectional light emanating from differential surface area $dA_1$ and divergent light from area $dA_2$. Although the radiation in region 1 is not a plane wave overall, it could represent a portion of a single spherical optical mode emitted by a point source located at the center of curvature, whereas the light issuing from a fixed $(x, y)$ location in region 2 is distributed over a range of angles $\theta$ and $\phi$ (and hence over many modes). The *curve* representing the overall surface $A$ could be the actual boundary of a sample emitting radiation, or it could simply be an arbitrary surface in space that the radiation happens to be crossing.

which should be carefully contrasted with (8.46). For narrowband light, this definition becomes

$$T_f \approx \frac{h\nu_0 \bar{n}}{k\left[(1+\bar{n})\ln(1+\bar{n}) - \bar{n}\ln\bar{n}\right]} \tag{8.49}$$

using (8.44) and (8.45). In accord with the discussion of (8.47) this flux temperature becomes infinite as $\bar{n} \to \infty$, consistent with the fact that an ideal laser beam carries zero entropy at a finite irradiance. Equations 8.49 and 8.16 are plotted in Figure 8.9 for 1-µm radiation, using (8.14) to relate the mean photon occupation number and spectral radiance. When $\bar{n} \ll 1$, these two temperatures become

$$\bar{T}_b \approx \frac{h\nu_0/k}{\ln(1/\bar{n})} \quad \text{and} \quad T_f \approx \frac{h\nu_0/k}{1 + \ln(1/\bar{n})}, \tag{8.50}$$

which leads to the simple formula

$$(kT_f)^{-1} \approx (k\bar{T}_b)^{-1} + (h\nu_0)^{-1}. \tag{8.51}$$

Using the values $\lambda_F = 995$ nm and $\bar{T}_b = 2000$ K for Yb$^{3+}$:ZBLAN given in connection with (8.23), this formula results in a fluorescence flux temperature of $T_f = 1750$ K. Note that (8.50) implies that $T_f \to \bar{T}_b$ as $\bar{n} \to 0$. In other words the flux and brightness temperatures become equal for sufficiently dim, narrowband radiation, as one can see in Figure 8.9. In contrast, $I_E^{BB} = \sigma T_b^4$ and $I_S^{BB} = \frac{4}{3}\sigma T_b^3$ for blackbody radiation (where $\sigma = 2\pi^5 k^4/15c^2 h^3$ is the Stefan–Boltzmann constant), so that $T_f = 0.75 T_b$ according to (8.48). This result states that if an isothermal body of unit emissivity radiates away heat $Q$ into free space, then while the object loses entropy $Q/T$, the thermal radiation carries away entropy $4Q/3T$. The net entropy change for this irreversible emission process in a zero-Kelvin environment is positive; entropy is only conserved if the surroundings are instead infinitesimally smaller in temperature than the body [15].

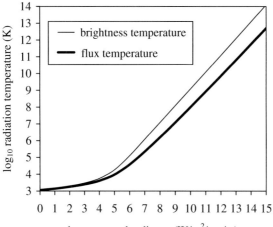

**Figure 8.9** Average brightness $\bar{T}_b$ and flux $T_f$ temperatures for narrowband infrared light peaking at $\lambda_0 = 1$ μm as a function of its mean spectral radiance $\bar{L}_\lambda$. The abscissa spans values ranging from low-power lamps up to high-brightness lasers. A radiation temperature is thus a useful, intensive figure of merit for evaluating the quality of an optical source at a given center wavelength; it characterizes the width of the distribution of energy among the optical modes.

## 8.4
### Ideal and Actual Performance of Optical Refrigerators

Consider the fluxes of energy and entropy into and out of the cooler sketched in Figure 8.10 operating at steady state. The first law of thermodynamics for this system becomes

$$\dot{E}_F = \dot{E}_P + \dot{Q}, \tag{8.52}$$

where $\dot{E}_P$ is here taken to be the absorbed pump power, after correcting for reflection, scattering, and transmission losses of the incident beam. Meanwhile, the second law states that

$$\dot{S}_F = \dot{S}_P + \frac{\dot{Q}}{T} + \dot{S}_G, \tag{8.53}$$

where $\dot{S}_G$ is the rate at which entropy is internally generated during a cooling cycle due to irreversible processes such as nonradiative relaxation, phonon equilibration, and spontaneous emission [34]. Substituting (8.48) into (8.53) leads to

$$\frac{\dot{E}_F}{T_F} = \frac{\dot{E}_P}{T_P} + \frac{\dot{Q}}{T} + \dot{S}_G = \frac{\dot{E}_P + \dot{Q}}{T_F} \tag{8.54}$$

using (8.52) to get the second equality, where $T_F$ and $T_P$ are the flux temperatures of the fluorescence and pump radiation, respectively. This result can be rearranged

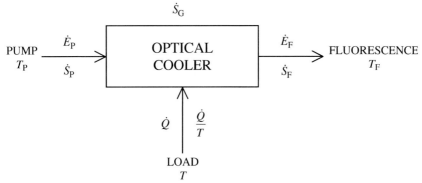

**Figure 8.10** Flow diagram analogous to Figure 8.1 for the rate at which energy and entropy are transported into and out of a fluorescent refrigerator or, in the case of entropy $\dot{S}_G$, is spontaneously generated during the cooling process. (Note that this latter entropy does not accumulate in the system at steady state because the entropy of the cooling sample, which is a function of state, must return to its initial value at the completion of each cycle. More entropy leaves than enters the system, as many of the processes occurring within the sample are irreversible.) To correspond to Figure 8.3, identify $\dot{E}_P$ with $\dot{E}_{in}$ and $\dot{E}_F$ with $\dot{E}_{out}$.

to compute the coefficient of performance (COP) $\kappa \equiv \dot{Q}/\dot{E}_P$, sometimes called the cooling efficiency relative to the absorbed pump power,

$$\kappa = \frac{T - \Delta \tilde{T}}{T_F - T} \quad \text{where} \quad \Delta \tilde{T} \equiv \frac{T T_F}{T_P}\left(1 + \frac{\dot{S}_G}{\dot{S}_P}\right) = \Delta T + T\frac{T_F \dot{S}_G}{\dot{E}_P} . \tag{8.55}$$

Notice that the maximum value for the COP is achieved for reversible operation of the cooler when $\dot{S}_G = 0$ so that $\Delta \tilde{T} = \Delta T$, thereby reproducing the Carnot value $\kappa_C$ of (8.4).

At the next level of approximation, assume that entropy is generated solely by nonradiative processes, such as direct multiphonon de-excitation of the active ions or energy transfer to nonfluorescent impurities. Define energy $E_F$ to be the fluorescent photon energy $h\nu_F$ multiplied by the number of excited ions in the sample $\mathcal{N}_2 = \int N_2 dV$, where $N_2$ is the population density of the upper manifold, as in (8.67) below. The radiative and nonradiative lifetimes of the excited state are $\tau_R$ and $\tau_{NR}$, respectively. Then the average fluorescent power is $\dot{E}_F = E_F/\tau_R$, which escapes from the sample, resulting in cooling. If that same amount of energy were to decay nonradiatively, then it would on average generate heating of $\dot{E}_H = E_F/\tau_{NR}$, which is deposited in the sample as thermal energy. (This is almost but not exactly equal to the actual nonradiative power which is $\dot{E}_{NR} = \mathcal{N}_2 h\nu_P/\tau_{NR}$. By energy conservation, each nonradiative relaxation must on average distribute a pump photon energy among the internal phonon modes, and thus the nonradiative energy per decay is $h\nu_P$, which only equals $h\nu_F$ if one pumps at the mean fluorescence wavelength.) The fluorescence quantum efficiency (QE) can now be written as

$$\eta \equiv \frac{1/\tau_R}{1/\tau_R + 1/\tau_{NR}} = \frac{\dot{E}_F}{\dot{E}_F + \dot{E}_H} \quad \Rightarrow \quad \dot{E}_H = \frac{1-\eta}{\eta}\dot{E}_F . \tag{8.56}$$

(Technically, the first equality defines the *internal* QE $\eta_{int} = \tau/\tau_R$, where $\tau$ is the overall lifetime; $\eta_{int}$ is only equal to the *external* QE $\eta_{ext}$ if the escape probability $f_{esc}$ is 100%. The correction for non-unit escape probability is discussed later.) Excess entropy results from the difference between dumping this thermal energy into the sample at temperature $T$ compared to carrying that energy away on the fluorescence beam at temperature $T_F$ [31],

$$\dot{S}_G = \frac{\dot{E}_H}{T} - \frac{\dot{E}_H}{T_F} = (\dot{E}_P + \dot{Q})\left(\frac{1-\eta}{\eta}\right)\left(\frac{1}{T} - \frac{1}{T_F}\right) \tag{8.57}$$

using (8.52) and (8.56) in the second step. Substituting this result into (8.55) now gives

$$\kappa = (T_F - T)^{-1}\left[T - \Delta T - TT_F(1+\kappa)\left(\frac{1-\eta}{\eta}\right)\left(\frac{1}{T} - \frac{1}{T_F}\right)\right]. \tag{8.58}$$

Equation 8.58 neatly simplifies to

$$\kappa = \eta\frac{T - \Delta T}{T_F - T} - 1 + \eta. \tag{8.59}$$

As a check, note that $\kappa$ reduces to $\kappa_C$ given by (8.4) when $\eta = 1$, because in this limit there is no nonradiative relaxation and hence no excess entropy generation. That Carnot expression for the COP can be used to rewrite (8.59) more compactly as

$$\kappa = \eta(1 + \kappa_C) - 1, \tag{8.60}$$

which is zero when $\eta \approx 1 - T/T_F = 83\%$ if $T = 300$ K and $T_F = 1750$ K $\gg \Delta T$ (as for Yb$^{3+}$:ZBLAN). Letting the fluorescence QE $\eta$ approach zero causes $\kappa = -1$ because 100% of the pump energy would then be converted into heat. More generally, solving (8.60) for $\kappa_C$ and substituting into it $\kappa = \dot{Q}/\dot{E}_P$ and $\eta = (\dot{E}_P + \dot{Q})/(\dot{E}_P + \dot{Q} + \dot{E}_H)$ from (8.56) using (8.52), one finds $\kappa_C = (\dot{Q} + \dot{E}_H)/\dot{E}_P$. This result makes sense because the numerator is the cooling power one would obtain if the nonradiative decays were radiative instead. Alternatively, one could start from this result and work backward to obtain (8.57), thereby proving that nonradiative relaxation is an irreversible source of entropy.

Another interpretation of (8.60) comes from substituting (8.9) into it in the form $1 + \kappa_C = \lambda_P/\lambda_F$. One then finds

$$\kappa = \frac{\lambda_P - \lambda_F^*}{\lambda_F^*}, \tag{8.61}$$

where $\lambda_F^* \equiv \lambda_F/\eta$ is the zero-heating wavelength corrected for nonradiative deexcitations. As expected intuitively, the effect of a non-unit fluorescence QE is to redshift further into the tail of the absorption band the minimum pump wavelength at which net cooling occurs.

Returning to the experimental cooling results for Yb$^{3+}$:ZBLAN, a titanium–sapphire pump laser is often used, which is continuous wave, narrowband and bright, so that $\bar{n} \gg 1$ in (8.49) and hence the pump flux temperature can be approximated as

$$T_P \approx \frac{hc\,\bar{n}_P}{k\lambda_P(1 + \ln \bar{n}_P)} \quad \text{with} \quad \bar{n}_P \approx \frac{\lambda_P^3 \dot{E}_P}{2hc\pi R_P^2 \Delta \nu_P \pi \delta_P^2}, \tag{8.62}$$

where the second equality comes from (8.14) and (8.17). Supposing the source has a power of $\dot{E}_P = 40$ W, a beam radius of $R_P = 0.5$ mm, a bandwidth of $\Delta \nu_P = 40$ GHz, a divergence of $\delta_P = 1$ mrad and a wavelength of $\lambda_P = 1030$ nm [15], then $\bar{n}_P \approx 10^9$ so that $T_P \approx 7 \times 10^{11}$ K. In comparison, the fluorescence flux temperature $T_F$ was computed above to be 1750 K using (8.51). Consequently, $\Delta T/T \equiv T_F/T_P \ll 1$ and (8.5) becomes an excellent estimate for the Carnot COP of a laser-driven optical cooler. Then $\kappa_C = (T_F/T - 1)^{-1}$ only depends on the ratio of the fluorescence flux and refrigerating sample temperatures. This result would also hold for an electroluminescent cooler, since an electric current delivers energy $E_P$ at near-zero entropy $S_P$ (ignoring inefficiencies such as Joule heating, Auger processes, and surface recombination), so that $T_P = E_P/S_P$ is again much larger than $T_F$.

So far we have only considered the interactions between the sample, pump, and fluorescence. However, the surroundings at ambient temperature $T_A$ (say the inner walls of the cryostat in which the sample is suspended by low-thermal-conductivity mounts) also couple radiatively to the refrigerator [18]. This coupling can be reduced for a practical cryocooler by inserting a set of heat shields between the sam-

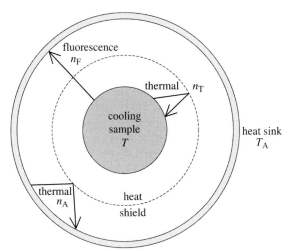

**Figure 8.11** Cylindrical cross-section of an optical cooler operating at temperature $T$. The blackened walls of the vacuum chamber are maintained at ambient temperature $T_A$ using an external coolant. The pump radiation is reflected between mirrors (not shown) parallel to the flat faces of the sample; one of these high reflectors could have a small input hole to admit fiber-coupled pump light [44]. The heat shield is assumed to have negligible emissivity (and thus absorptivity), reflecting all thermal radiation at long wavelengths and transmitting all fluorescence at short wavelengths. The sample reflects or transmits some of the incident thermal flux with occupation number $n_T$; it absorbs and then re-emits the remaining portion.

ple and the walls, as in Figure 8.11. For example, for an ytterbium-based cooler mounted in a room-temperature vacuum chamber, the heat shield would be designed to transmit the 1-µm fluorescence out to the heat-sunk walls while reflecting their 10-µm thermal radiation (with corresponding black-body photon occupation number $n_A$) back away from the sample. The heat shield is thus a short-wavelength-pass filter (sometimes called a "hot mirror"). Since the cooler is presumably operating at a temperature $T$ less than $T_A$, thermal radiation from the sample (with occupation number $n_T$) will peak at wavelengths even longer than 10 µm and will consequently be reflected back to the sample. Ignoring the pump, it therefore follows that the light incident on the cooling sample will have occupation number $n_T$ while that leaving it will have occupation number $n_F + n_T$. The net energy and entropy fluxes from the sample result from the difference between the outgoing and incoming radiation,

$$\dot{E}_{net} = 2hc^{-2} \int_A \int_\Omega \int_\nu (n_F + n_T) \nu^3 \, d\nu \cos\theta \, d\Omega \, dA$$
$$- 2hc^{-2} \int_A \int_\Omega \int_\nu n_T \nu^3 \, d\nu \cos\theta \, d\Omega \, dA , \qquad (8.63)$$

which simplifies back to (8.13), with $n$ identified as $n_F$, while

$$\dot{S}_{net} = 2kc^{-2} \int_A \int_\Omega \int_\nu \left[ (1 + n_F + n_T)\ln(1 + n_F + n_T) - (n_F + n_T)\ln(n_F + n_T) \right]$$
$$\times \nu^2 \, d\nu \cos\theta \, d\Omega \, dA$$
$$- 2kc^{-2} \int_A \int_\Omega \int_\nu \left[ (1 + n_T)\ln(1 + n_T) - n_T \ln n_T \right] \nu^2 \, d\nu \cos\theta \, d\Omega \, dA$$
$$(8.64)$$

does *not* reduce to (8.43) when $n$ is similarly identified. Unlike energy, the entropy of a beam of light is not additive over photons that share the same mode (in particular, the same frequency). This fact means that the entropy of the sample's emission depends on the temperature of the thermal surroundings. Specifically, in the limit of very weak fluorescence $n_F \ll n_T$, we can neglect $n_F$ in the arguments of the first two logarithms in (8.64) to get

$$\dot{S}_{net} \approx 2kc^{-2} \int_A \int_\Omega \int_\nu n_F \ln(1 + 1/n_T) \nu^2 \, d\nu \cos\theta \, d\Omega \, dA . \qquad (8.65)$$

However, $\ln(1 + 1/n_T) = h\nu/kT$ from (8.39) and thus $\dot{S}_{net} = \dot{E}_{net}/T \Rightarrow T_F \to T$. This result keeps the Carnot coefficient of performance in (8.5) positive regardless of how weak the fluorescence gets. In fact, as the pump power (and hence the resulting fluorescence) is reduced, $\kappa_C$ increases, as graphed by the left-hand curve in Figure 8.12, diverging as $\dot{E}_P \to 0$. The reason the COP becomes infinite in this limit is that the sample will still cool back to the ambient temperature if it is impulsively heated, so that there is a nonzero cooling power even though the pump power is zero, owing to the sample's radiative coupling to the surroundings (arising from a small leakage of thermal radiation through the heat shield in Figure 8.11 at short

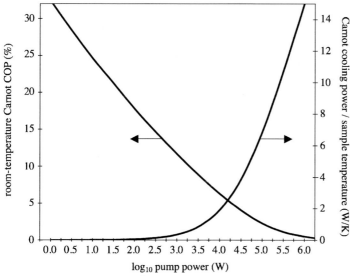

**Figure 8.12** Plots of $\kappa_C$ and of $\dot{Q}/T$ (in the Carnot limit) for pump powers $\dot{E}_P$ ranging from 1 W to 1 MW. As in Section 8.3, the cooling sample is assumed to have a surface area of $A_{sample} = 13.5\pi \text{ cm}^2$ and the fluorescence spectrum is that of Yb$^{3+}$:ZBLAN at room temperature, peaking at $\lambda_F = 995$ nm with a bandwidth of $\Delta\lambda_F = 35$ nm. The Carnot COP is computed at $T = 300$ K, assuming $T_P \gg T_F$.

wavelengths). One can equivalently think of this as thermally stimulated fluorescence. Needless to say, the fact that $\kappa_C$ increases with decreasing $\dot{E}_P$ does not imply that there is some maximum in the cooling power of an ideal optical refrigerator at low or intermediate pump powers. On the contrary, for a cooler operating at the Carnot limit, (8.53) implies that $\dot{Q}/T = \dot{S}_F$ (assuming the entropy of the pump is negligible), which is plotted as the right-hand curve in Figure 8.12 for narrowband ytterbium fluorescence and rises monotonically with pump power.

Equation 8.65 is only valid if the fluorescence is extremely weak. Specifically, when $T = 300$ K and $c/\nu = 995$ nm, (8.39) gives $n_T = 10^{-21}$. Using the previous values of the material constants, (8.23) implies that $\bar{n}_F = n_T$ only when the fluorescence power has the tiny value $\dot{E}_F = 10^{-16}$ W. Clearly we can assume that $n_F \gg n_T$ within the ytterbium spectral range whenever the optical fridge is operating. In this limit of small $n_T$, the quantity in square brackets in the second integral of (8.64) becomes approximately $(-\ln n_T)/(1/n_T) \approx 0$ using l'Hôpital's rule. Therefore, (8.64) now *does* simplify to (8.43), with $n$ identified as $n_F$.

With that conclusion in mind, the graphs in Figure 8.12 were computed as follows. A value of the fluorescence occupation number $\bar{n}_F$ was picked and $\dot{E}_F$ was computed from it using (8.23). Similarly, the fluorescence entropy flux was calculated as

$$\dot{S}_F = 2\pi k c A_{sample} \left[(1 + \bar{n}_F) \ln(1 + \bar{n}_F) - \bar{n}_F \ln \bar{n}_F\right] \lambda_F^{-4} \Delta\lambda_F \tag{8.66}$$

from (8.45). Then the fluorescence flux temperature was found from their ratio, $T_F \equiv \dot{E}_F/\dot{S}_F$, and it was substituted into (8.5) to determine $\kappa_C$. Finally the pump power could be determined as $\dot{E}_P = \dot{E}_F/(1+\kappa_C)$ using (8.52), so that $\kappa_C$ and $\dot{S}_F$ could be plotted one point at a time against $\dot{E}_P$. Unlike $\kappa_C$, whose value depends critically on the sample temperature $T$, assumed to be room temperature in Figure 8.12, $\dot{Q}/T$ as given by (8.66) only depends weakly on $T$, via the temperature dependences of $\lambda_F$ and $\Delta\lambda_F$ measured by Lei et al. [35], noting from the two graphs in Figure 8.12 that $\dot{E}_P \approx \dot{E}_F$ when the pump is strong enough to give significant cooling. Also note that the right-hand plot of $\dot{Q}/T$ versus $\log \dot{E}_P$ becomes linear at high powers because then $\dot{S}_F \propto \ln \bar{n}_F$ from (8.66) while $\dot{E}_P \approx \dot{E}_F \propto \bar{n}_F$.

The current lowest temperature experimentally attained by optical cooling is $T = 208$ K for a 2 wt % $Yb^{3+}$:ZBLAN cylindrical sample starting from room temperature [36]. The mean fluorescence wavelength at this temperature is $\lambda_F = 999$ nm and the bandwidth is approximately the same as its room temperature value of $\Delta\lambda_F = 35$ nm. Since the end faces of the sample were coated with high reflectors that substantially reduce fluorescence escape from them, it is reasonable to estimate the sample area only by that of the curved surface, specifically of the outer cladding (since $\delta = \pi/2$ for it), so that $A_{\text{sample}} = 88\pi \text{ mm}^2$. The sample was pumped using an $Yb^{3+}$:YAG laser with an absorbed power of $\dot{E}_P = 5.9$ W $\approx \dot{E}_F$ (assuming 61% absorptance) at a wavelength of $\lambda_P = 1026$ nm for 3 h (to reach steady state). Consequently (8.23) implies that the mean fluorescence occupation number is $\bar{n}_F = 1.6 \times 10^{-3}$, corresponding to a flux temperature of $T_F = 1900$ K from (8.50). The sample and flux temperatures then lead to a Carnot COP of $\kappa_C = 12\%$ according to (8.5). For comparison, the coefficient of performance predicted spectroscopically is $\kappa_C = 2.7\%$ from (8.9). However, the heat load at the minimum temperature was found to be $\dot{Q} = 29$ mW, and thus the actual COP is just $\kappa = \dot{Q}/\dot{E}_P = 0.5\%$.

Another sample for which more than 10 °C of cooling has been observed is $Tm^{3+}$:ZBLAN [37,38]. A Brewster-cut 1 wt % sample was cooled by 24 K below room temperature (i.e. to about $T = 275$ K) by exciting it at $\lambda_P = 1.9$ μm for half an hour with an average absorbed power of $\dot{E}_P = 2.2$ W $\approx \dot{E}_F$ using an optical parametric oscillator pumped by a 25-W mode-locked $Nd^{3+}$:YAG laser. The surface area of the sample was about $A_{\text{sample}} = 150 \text{ mm}^2$. The fluorescence spectrum consists of a single symmetric peak centered at $\lambda_F = 1.803$ μm with a FWHM of $\Delta\lambda_F = 0.22$ μm. Using these values, one finds that $\bar{n}_F = 3.4 \times 10^{-3}$ from (8.23), so that $T_F = 1200$ K according to (8.50). Therefore, (8.5) gives rise to $\kappa_C = 30\%$, whereas the ideal spectroscopic prediction is $\kappa_C = 5.4\%$ according to (8.9). Note that the latter value is double that calculated for $Yb^{3+}$:ZBLAN above because, as discussed following (8.9), $\kappa_C$ is approximately proportional to $\lambda_P$ and for thulium we double this wavelength from about 1 to 2 μm. With an estimated cooling power of $\dot{Q} = 73$ mW, the actual COP may be as high as $\kappa = \dot{Q}/\dot{E}_P = 3.3\%$, which is promising for practical refrigeration.

To explain the differences between the thermodynamic, spectroscopic, and actual cooling coefficients of performance, we need to consider various sources of inefficiencies. Equation 8.61 for example indicates that $\kappa$ strongly depends on the internal fluorescence quantum efficiency $\eta_{\text{int}}$. In addition, the COP is expected to depend on the probability $f_{esc}$ that the spontaneously emitted photons escape from the sample, the background absorption coefficient $\alpha_{back}$ of the pump light by nonfluorescent impurities or sample surface coatings, and the saturation intensity (irradiance) $I_{sat}$ when one pumps near an absorption peak. The following model, which extends that of Hoyt et al. [37], incorporates all four of these effects ($\eta_{int}$, $f_{esc}$, $\alpha_{back}$, and $I_{sat}$).

Consider a small volume element of the sample $dV$ at spatial position $\mathbf{r}$ that is being optically pumped with incident intensity $I(\mathbf{r})$ in W/cm² at wavelength $\lambda_P = c/\nu_P$. The population densities (ions/cm³) in the ground and excited manifolds are $N_1$ and $N_2$, respectively, where $N = N_1 + N_2$ is the concentration of active ions doped into the host crystal or glass and is assumed to be spatially uniform. Denote the effective absorption and emission cross-sections (in cm²) at the pump wavelength as $\sigma_{AP}$ and $\sigma_{EP}$, respectively, and define [39] the dimensionless ratio $\beta \equiv \sigma_{AP}/(\sigma_{AP} + \sigma_{EP})$. (Note that $\beta = 0.5$ for a true two-level system.) The radiative and nonradiative rates are the inverses of the lifetimes, $W_R = 1/\tau_R$ and $W_{NR} = 1/\tau_{NR}$, respectively.

At steady state, a rate-equation approach leads to two key relations. First, the time dependence of the excited-state population is described by

$$\frac{dN_2}{dt} = 0 = \frac{I}{h\nu_P}(N_1 \sigma_{AP} - N_2 \sigma_{EP}) - f_{esc} N_2 W_R - N_2 W_{NR}, \tag{8.67}$$

assuming the pump bandwidth $\Delta\lambda_P$ is narrow enough that the wavelength dependence of the cross-sections can be ignored; otherwise one needs to replace each product $I\sigma$ by $I_\lambda \sigma(\lambda)$ and integrate (8.67) and (8.71) over wavelength. The first expression on the right-hand side of (8.67) is the difference between absorption and stimulated emission, the second term describes spontaneous emission, and the last one accounts for nonradiative decay (including both direct multiphonon deexcitation and energy transfer from the excited ions to nonfluorescent impurities). The spontaneous radiation term includes the fractional probability $f_{esc}$ that the fluorescence photons ultimately escape from the sample. Photons that do not escape are assumed to get reabsorbed by active ions (i.e. perfect photon recycling), resulting in no net change in the excited-state population; Wang et al. [40] have considered the case of nonideal recycling and its effect is to further reduce the external quantum efficiency. For simplicity, here $f_{esc}$ is taken to be an average value over the entire sample; in actuality it depends on $\mathbf{r}$ both because of the proximity of the pumped volume element to the sample surfaces and because the photons emitted in one volume element are in general absorbed in a different volume element [41]. In contrast, the stimulated emission photons are assumed to be added to the pump beam, and so no escape fraction is needed inside the first term. Define the external fluorescence quantum efficiency (QE) as

$$\eta_{ext} \equiv \frac{f_{esc} W_R}{f_{esc} W_R + W_{NR}}, \tag{8.68}$$

which can be related to the internal QE $\eta_{int}$ defined by the first equality in (8.56),

$$\eta_{ext}^{-1} - 1 = \frac{\eta_{int}^{-1} - 1}{f_{esc}}. \tag{8.69}$$

This expression can be used to directly compute the external QE in terms of the internal value (or vice versa) if the escape probability is known. (Note that if $f_{esc} \neq 1$, the internal and exernal efficiencies are only equal to each other in the limits that $\eta$ approaches zero or unity.) Using (8.68) to eliminate $W_{NR}$ in (8.67), one obtains

$$\frac{N_2}{N} = \frac{\beta}{1 + I_{sat}/I}, \tag{8.70}$$

where the pump saturation intensity is $I_{sat} \equiv h\nu_P f_{esc} W_R / \eta_{ext}(\sigma_{AP} + \sigma_{EP})$.

The second key relation is the rate of thermal energy accumulation in the volume element,

$$\dot{u} = (N_1 \sigma_{AP} - N_2 \sigma_{EP})I + \alpha_{back} I - f_{esc} N_2 h\nu_F W_R \tag{8.71}$$

in W/cm$^3$, where $\alpha_{back}$ is an average background nonsaturable absorption coefficient (in cm$^{-1}$) which is assumed to be approximately wavelength independent due to nonfluorescent impurities and surface coatings, and $\nu_F = c/\lambda_F$ is the mean fluorescence frequency including the redshifting due to reabsorption. The redshift can be calculated from the overlap between the absorption and emission spectra; for example, Lamouche et al. [42] estimate a +9 nm shift for a 2% Yb$^{3+}$:ZBLAN sample measuring 3 cm on a side. Noting that the resonant absorption coefficient by the active ions is $\alpha_{res} = N\sigma_{AP}$, and defining the total absorption coefficient as $\alpha_{tot} = \alpha_{res} + \alpha_{back}$, we can use (8.70) to rewrite (8.71) in dimensionless form as

$$\frac{\dot{u}}{\alpha_{tot} I} = 1 - \frac{\eta_{ext} \lambda_P / \lambda_F + I/I_{sat}}{1 + I/I_{sat}} \frac{\alpha_{res}}{\alpha_{tot}}. \tag{8.72}$$

The numerator on the left-hand side is the net heating power density while the denominator is the total absorbed power density. Therefore, the negative of this expression defines the cooling COP, which can be rewritten in the form

$$\kappa = \frac{\lambda_P - \lambda_F^*}{\lambda_F^{**}}. \tag{8.73}$$

Here the zero-heating wavelengths are

$$\lambda_F^* = \frac{\lambda_F}{\eta_{ext}} \left( \frac{\alpha_{tot} + \alpha_{back} I/I_{sat}}{\alpha_{res}} \right), \tag{8.74}$$

while the slope of a graph of $\kappa$ versus $\lambda_P$ (as in Figure 8.14 introduced below) is normalized by

$$\lambda_F^{**} = \frac{\lambda_F}{\eta_{ext}} \left( 1 + \frac{I}{I_{sat}} \right) \frac{\alpha_{tot}}{\alpha_{res}}. \tag{8.75}$$

As a check, (8.73) reduces to (8.61) when $f_{esc} = 1$ (so that the internal and external quantum efficiencies are equal), $\alpha_{back} = 0$ (so that the pump light is only absorbed by the active ions), and $I_{sat} \to \infty$ (so that the absorption by the active ions cannot be saturated). More realistically, $I_{sat} = 15$ kW/cm$^2$ if Yb$^{3+}$:ZBLAN is pumped at $\lambda_P = 1$ μm, so that $\sigma_{AP} + \sigma_{EP} \approx 5 \times 10^{-21}$ cm$^2$ at room temperature, assuming values of $f_{esc} = 75\%$, $\tau_R = 2$ ms, and $\eta_{ext} = 99\%$. This irradiance would be attained

## 8.4 Ideal and Actual Performance of Optical Refrigerators

by a 1-W pump beam with a diameter of about 0.1 mm, and therefore it is not surprising that one sees saturation effects when performing focused photothermal deflection spectroscopic measurements or when the cooling sample is a piece of an optical fiber [34]. In either case, the result is to reduce the amount of laser heating or cooling observed, since both $\lambda_F^*$ and $\lambda_F^{**}$ increase with increasing pump intensity $I$ in (8.74) and (8.75). As an example, the ratio of the COP to its unsaturated value has been plotted in Figure 8.13 as a function of the pump intensity for the case of $\alpha_{back} = 0$.

Due to such saturation effects, the right-hand curve of the cooling power in Figure 8.12 must roll over at high pump powers, rather than continuing to increase without limit, although a "top hat" spatial profile for the laser beam can help delay that onset. In any event, the cooling rate is ultimately limited to $1/\tau_R = 500$ Hz per ytterbium ion.

On the other hand, in the limit of weak pumping ($I \ll I_{sat}$), which will always occur at sufficiently long wavelengths, (8.74) and (8.75) become $\lambda_F^* = \lambda_F^{**} = \lambda_F/(\gamma \eta_{ext})$, where the absorption efficiency [43] is $\gamma \equiv \alpha_{res}/\alpha_{tot}$. The resulting expression for $\kappa$ from (8.73) has been plotted in Figure 8.14 using values of the numerical constants representative of purified $Yb^{3+}$:ZBLAN material, assuming a sample geometry such that fluorescence reabsorption is negligible and estimating the ytterbium absorption coefficient as

$$\alpha_{res}(\lambda_P) = (0.36 \text{ cm}^{-1}) \exp\left[-\left(\frac{\lambda_P - 975 \text{ nm}}{33 \text{ nm}}\right)^2\right], \tag{8.76}$$

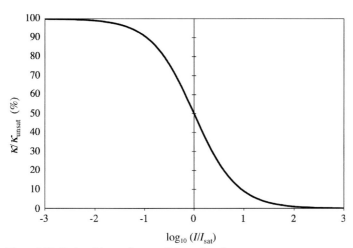

**Figure 8.13** Ratio of the cooling coefficient of performance $\kappa$ to its unsaturated value $\kappa_{unsat} \equiv \eta_{ext}\lambda_P/\lambda_F - 1$ plotted against pump intensity $I$ (varying over a range from a thousandth to a thousand times the saturation intensity $I_{sat}$). Note in particular that the COP is halved when one drives the cooler at $I_{sat}$, because at saturation half of the emissions are stimulated rather than spontaneous and so do not contribute to cooling.

which was fitted to the 300 K reciprocity-derived spectrum of a 1% doped sample over the range 990–1060 nm [35]. We see from this graph that the cooling range is bracketed by two zero-heating wavelengths. At short pump wavelengths, $\alpha_{\text{back}}$ is negligible compared to $\alpha_{\text{res}}$ and (8.73) becomes

$$\kappa = \frac{\lambda_P - \lambda_F/\eta_{\text{ext}}}{\lambda_F/\eta_{\text{ext}}} = \frac{\dot{E}_F}{\dot{E}_F \lambda_F/\lambda_P + \dot{E}_{\text{NR}}} - 1. \tag{8.77}$$

The first equality is identical to (8.61) with the internal fluorescence QE replaced by the external QE, and thus $\kappa$ rises linearly with $\lambda_P$ and crosses zero near $\lambda_F/\eta_{\text{ext}} = 1005$ nm. The second equality follows by substituting (8.68) and defining the external fluorescence power by $\dot{E}_F = f_{\text{esc}} \mathcal{N}_2 h\nu_F/\tau_R$ and the nonradiative heating by $\dot{E}_{\text{NR}} = \mathcal{N}_2 h\nu_P/\tau_{\text{NR}}$, as discussed before (8.56). Some quick checks on any purported expression for the cooling coefficient $\kappa$ (of which there have been many in the literature) are that $\kappa = -1$ if $\tau_R \to \infty$ or if $f_{\text{esc}} = 0$ (in which case the best you can do is locally cool one region of the sample and distribute the heat elsewhere, assuming the material's thermal conductivity is low enough to support such a gradient), and that $\kappa = \kappa_C$ from (8.9) in the absence of saturation and all nonradiative heating (such as from background or excited-state absorption, multiphonon relaxation, and energy transfer). Equation 8.77 satisfies these checks when one sets $\dot{E}_F$ or $\dot{E}_{\text{NR}}$ to zero, respectively. Note that the denominator of the second equality can be written as $f_{\text{esc}} \mathcal{N}_2 h\nu_P/\tau_R + \mathcal{N}_2 h\nu_P/\tau_{\text{NR}} = \mathcal{N}_2 h\nu_P/\tau_{\text{ext}}$, where $\tau_{\text{ext}}$ is the external lifetime of the active ions, which is longer than the the internal value owing to radiative trapping of the fluorescence. However, $\mathcal{N}_2 h\nu_P/\tau_{\text{ext}}$ equals the absorbed pump power $\dot{E}_P$ at steady state. Equation 8.77 can then be immediately recognized as the expected ratio of the cooling to the absorbed power.

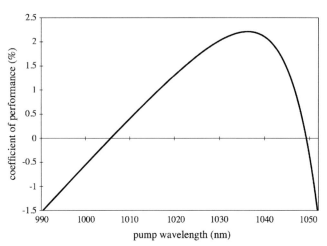

**Figure 8.14** Plot of the COP $\kappa$ versus the pump wavelength $\lambda_P$ for 1 wt % Yb$^{3+}$:ZBLAN at room temperature assuming a mean fluorescence wavelength of $\lambda_F = 995$ nm, an external fluorescence quantum efficiency of $\eta_{\text{ext}} = 99\%$, and a background absorption coefficient of $\alpha_{\text{back}} = 10^{-4}$ cm$^{-1}$. The pump intensity is presumed to be much weaker than the saturation intensity, $I \ll I_{\text{sat}}$, at all wavelengths longer than 990 nm.

On the other hand, as one tunes to long wavelengths, $\alpha_{res}(\lambda_P)$ declines in value while $\alpha_{back}$ is assumed constant, so that the cooling COP does not continue to increase but instead bends over, returning to zero when $\alpha_{res}(\lambda_P) = \alpha_{back}/(\eta_{ext}\lambda_P/\lambda_F - 1)$. Substituting (8.76), this second zero-heating pump wavelength is found at 1049 nm. The peak cooling in Figure 8.14 occurs near $\lambda_P = 1036$ nm, in reasonable agreement with experimental results [44].

## 8.5 Closing Remarks

This chapter has focused on calculating the coefficient of performance (COP) $\kappa$. Many workers in the field of optical cooling prefer to report quantities related to the COP rather than the COP itself. The cooling power $P_{cool}$ equals the product of $\kappa$ and the absorbed pump power $P_{abs} = \dot{E}_P$. In the unsaturated limit, the absorbed power in turn is the product of the incident power $P_{inc}$ and the absorptance $1 - \exp(-\alpha_{res}L)$, where $L$ is the sample length (multiplied by the number of passes if it is in a cavity). For a short sample in single-pass pumping, one can therefore approximate $P_{abs} \approx \alpha_{res}LP_{inc}$. Once the steady-state refrigeration temperature $T$ has been attained, the cooling power is equal to the heat load, $P_{cool} = \dot{Q}$. If the load is thermal radiation from a surrounding chamber (possibly coated with a heat-shielding material) of internal surface area $A_c$ at temperature $T_c$ with an emissivity of $\varepsilon_c$, then [45]

$$\dot{Q} = \frac{\sigma(T_c^4 - T^4)}{(A\varepsilon)^{-1} + (1 - \varepsilon_c)(A_c\varepsilon_c)^{-1}}, \tag{8.78}$$

where the cooling sample has surface area $A$ and emissivity $\varepsilon$. An upper limit on the heat load is obtained by putting $\varepsilon_c = 1 = \varepsilon$. Furthermore, if the cooler is operating at a temperature $T = T_c - \delta T$ where $\delta T$ is small, then $\dot{Q} \approx 4A\sigma T_c^3 \delta T$. Suppose the sample is in the form of an optical fiber of small diameter $D$, so that $A \approx \pi DL$. Then the COP becomes

$$\kappa \equiv \frac{\dot{Q}}{P_{abs}} \approx \frac{4\pi\sigma DT_c^3 \delta T}{\alpha_{res} P_{inc}} \tag{8.79}$$

so that, ideally using (8.9), the temperature drop normalized to the incident laser power is

$$\frac{\delta T}{P_{inc}} = \frac{N}{4\pi\sigma DT_c^3} F_{cool}, \tag{8.80}$$

where $N$ is the concentration of active ions and the cooling figure of merit [39] is $F_{cool} \equiv \sigma_{AP}(\lambda_P - \lambda_F)/\lambda_F$ at the optimal pump wavelength $\lambda_P$, which is useful for comparing the cooling potentials of different materials. Equation 8.80 suggests possible strategies for maximizing the temperature drop, such as increasing the $Yb^{3+}$ doping (until energy transfer becomes limiting) and decreasing the sample diameter (so that its entire cross-section is pumped).

From a spectroscopic point of view, the ratio of the external radiative relaxation rate to the total decay rate of the upper state defines the fluorescence quantum efficiency (QE),

$$\eta_{\text{ext}} \equiv \frac{f_{\text{esc}}/\tau_R}{1/\tau_{\text{ext}}}, \tag{8.81}$$

where $1/\tau_{\text{ext}} \equiv f_{\text{esc}}/\tau_R + 1/\tau_{\text{NR}}$. However, the ratio of the external fluorescent power $\dot{E}_F$ to the absorbed pump power defines the cooling coefficient of performance (COP) plus unity,

$$\kappa + 1 \equiv \frac{f_{\text{esc}} N_2 h\nu_F/\tau_R}{N_2 h\nu_P/\tau_{\text{ext}}}, \tag{8.82}$$

in the absence of both saturation and background absorption. (Note that photon recycling will be perfect in this case because nothing other than active ions can reabsorb the fluorescent light.) Thus, $\kappa + 1 = \eta_{\text{ext}} \nu_F/\nu_P$, which is (8.60), noting that $\nu_F/\nu_P = \kappa_C + 1$ from (8.9).

On the other hand, from a thermodynamic viewpoint, the Carnot COP is $\kappa_C = T/(T_F - T)$ provided that $T_P \gg T_F$. Here the flux temperature for narrowband fluorescence is

$$T_F = \frac{h\nu_F \bar{n}_F}{k\left[(1+\bar{n}_F)\ln(1+\bar{n}_F) - \bar{n}_F \ln \bar{n}_F\right]} \tag{8.83}$$

from (8.49), where the mean photon occupation number is given by (8.23) as

$$\bar{n}_F = \frac{c^2 \dot{E}_F}{2\pi h \nu_F^3 \Delta\nu_F A_{\text{sample}}} \tag{8.84}$$

and $\dot{E}_F = (1+\kappa) P_{\text{abs}} \approx P_{\text{abs}}$ according to (8.82). Consequently, the parameters needed to calculate the ideal COP are the absorbed pump power $P_{\text{abs}}$, the sample's cooled temperature $T$ and uncoated surface area $A_{\text{sample}}$, and the center frequency $\nu_F$ and bandwidth $\Delta\nu_F$ of the fluorescence spectrum (at temperature $T$). It would therefore be useful if experimentalists made it a standard practice to cite values for these five parameters in their papers.

How practical would it be to optically pump the cooler with a *nonlasing* source of light? Equation 8.83 gives the flux temperature $T_P$ of a narrowband pump if we replace $\nu_F$ by $\nu_P$ and $\bar{n}_F$ by $\bar{n}_P$ where

$$\bar{n}_P = \frac{c^2 \dot{E}_P}{2 h \nu_P^3 \Delta\nu_P A_{\text{pump}} \pi \sin^2 \delta_P}, \tag{8.85}$$

for a pump beam of bandwidth $\Delta\nu_P$, divergence $\delta_P$, and cross-sectional area $A_{\text{pump}}$. Noting from (8.4) that

$$\kappa_C = \frac{1 - \frac{T_F}{T_P}}{\frac{T_F}{T} - 1}, \tag{8.86}$$

little reduction in the Carnot COP will result as long as one maintains $T_P \gg T_F$. However, $\dot{E}_F \approx \dot{E}_P$ and $\nu_F \approx \nu_P$, so that this temperature requirement becomes

$$\Delta\nu_P A_{pump} \sin^2 \delta_P \ll \Delta\nu_F A_{sample}. \tag{8.87}$$

Since the fluorescence and absorption bandwidths of a sample are comparable, satisfying this inequality guarantees that the pump is strongly absorbed. However, strong absorption is a necessary but not sufficient condition to obtain cooling. For example, if we simply reflected the fluorescence back to the sample (either accidentally from the walls of the sample chamber or intentionally in a misguided attempt to "recycle" the fluorescence energy) and called it "pump" light, then the left-hand side of (8.87) would be roughly equal to rather than much smaller than the right-hand side; that is, $T_P \approx T_F$ for that portion of the input radiation. Under ideal conditions, that would leave the COP unchanged, as we see from (8.86); but in reality it would *decrease* the COP according to (8.55), owing to background absorption, energy transfer to nonfluorescent impurities, and other heating inefficiencies.

Frey et al. [46] suggest downshifting (and, implicitly, frequency narrowing and spatially collimating) the fluorescence before recycling it. More practically, one could use photovoltaics to convert some of the fluorescence into electrical energy and use that to help run the pump source. The highest possible efficiency results when the set of (uncooled) photovoltaic converters constitute a Carnot heat engine operating between the fluorescence flux temperature $T_F$ and room temperature $T_R$, so that

$$\varepsilon_C = \frac{T_F - T_R}{T_F} \tag{8.88}$$

as in (8.3). This efficiency is the ratio of the converted electrical power $\dot{E}_R$ (to be recycled back perfectly to the pump power $\dot{E}_P$ input to the optical cooler) and the collected fraction $\chi$ of the cooler's fluorescence power $\dot{E}_F$, $\varepsilon_C \equiv \dot{E}_R/\chi\dot{E}_F$. The Carnot COP of the coupled refrigerator–photovoltaic system now becomes the ratio of the cooling to the net electrical work supplied,

$$\kappa_{PV} \equiv \frac{\dot{Q}}{\dot{E}_P - \dot{E}_R} = \frac{T}{T^* - T}, \quad \text{where} \quad T^* \equiv \chi T_R + (1-\chi)T_F, \tag{8.89}$$

since $\dot{E}_F = \dot{E}_P + \dot{Q}$ and $\dot{Q} = \dot{E}_P T/(T_F - T)$. As a check, (8.89) reduces to (8.5) if $\chi = 0$ so that there is no fluorescent recycling. On the other hand, for perfect collection ($\chi = 1$) of the fluorescence, $T_F$ is replaced by $T_R$ in (8.5), which amounts to a significant improvement in efficiency, because $T_F > T_R$. In effect, we are then exhausting waste heat out of the refrigerator system at room temperature $T_R$ rather than at the fluorescence temperature $T_F$. More generally the ratio of the COP with the photovoltaic recycler to that of (8.5) in its absence is

$$\frac{\kappa_{PV}}{\kappa_{NPV}} = \left[1 - \chi \frac{T_F - T_R}{T_F - T}\right]^{-1}. \tag{8.90}$$

For example, if we use the values calculated after (8.66) for the current best Yb$^{3+}$:ZBLAN cooler ($T$ = 208 K and $T_F$ = 1900 K) assuming $T_R$ = 300 K and say $\chi$ = 50%, then (8.90) predicts that one can almost double the ideal cooling performance.

In another vein, an intriguing application of fluorescent cooling is to reduce the thermal load on the medium in an optically pumped laser [47]. Such heating results in deleterious effects such as beam defocusing by thermal lensing, depolarization due to temperature-dependent birefringence, coating delamination, and stress fractures. To reduce these effects, high-power solid-state lasers are typically water cooled. Not only does this add bulk to the entire system, it does not fully eliminate the problems, because only the surface of the laser rod or slab is being directly cooled while it is the interior that is pumped, and thus there is a thermal gradient which ultimately limits the extent to which one can scale up the power. In contrast, optical cooling occurs inside the medium itself. One could start by imagining two separate systems: one pump source and set of active ions to drive the laser, and a second pump source and set of ions to run the optical cooler. Essentially this idea has been proposed by Petrushkin and Samartsev [48], in which a KY$_3$F$_{10}$ crystal is double-doped with Nd$^{3+}$ and Yb$^{3+}$. Suppose the neodymium ions lase at a wavelength $\lambda_L$ following optical pumping at a wavelength $\lambda_P$. The Stokes energy shift $hc(\lambda_P^{-1} - \lambda_L^{-1})$ is called the quantum defect and heats the crystal. However, some fraction (determined by the doping concentrations) of the laser photons are absorbed in the long-wavelength tail of the ytterbium spectrum and consequently promote anti-Stokes fluorescent cooling of $hc(\lambda_F^{-1} - \lambda_L^{-1})$ per cycle. In principle, one could balance the cooling against the heating, resulting in what has been termed *athermal* laser operation. However, this particular scheme for high-power scaling is limited because laser photons are being consumed by the cooling process and the laser wavelength has to be chosen to overlap the absorption band of the cooling ions and not just according to the optimal emission of the lasing ions.

A clever alternative dispenses with the need for a separate optical cooling system and uses a single pumped set of active ions which both lase by Stokes-shifted stimulated emission and fluoresce by anti-Stokes-shifted spontaneous emission. This is done by choosing a pump photon frequency which is intermediate between the mean fluorescence frequency and the laser output frequency (selected by appropriate design of the feedback cavity mirrors), $\nu_L < \nu_P < \nu_F$. Using a biaxial host such as KGd(WO$_4$)$_2$, one can choose the pump wavelength and polarization to coincide with an absorption peak and independently choose the laser wavelength and polarization to match an emission peak. Athermal operation occurs when two balance conditions are met in each volume element of the laser medium: both the rates and the powers for absorption and (spontaneous plus stimulated) emission must be equal, using equations similar to those of (8.67) and (8.71). Such a laser is said to be *radiation balanced*. A combined optical and thermodynamic analysis of a single-pass radiation-balanced Yb$^{3+}$:KGd(WO$_4$)$_2$ amplifier has been undertaken by Mungan [29], and its Carnot efficiency for the ratio of the output laser power to the input pump power is $\varepsilon_C \equiv \dot{E}_L/\dot{E}_P = 1 - T_F/T_P$, since ideally the laser is a heat engine operating between the pump and fluorescence flux temperatures. An oscillator model has been discussed by Li et al. [49].

Consider the thermal load on a continuous-wave (cw) laser medium in the ideal case where there is no reabsorption, nonradiative relaxation, or background absorption. At steady state, the rate of optical pumping (transitions per second) up to the excited state must balance the relaxation rate back down to the ground state by lasing and fluorescence,

$$\dot{\mathcal{N}}_P = \dot{\mathcal{N}}_L + \dot{\mathcal{N}}_F \quad \Rightarrow \quad \frac{\dot{E}_F}{h\nu_F} = \frac{\dot{E}_P}{h\nu_P} - \frac{\dot{E}_L}{h\nu_L}. \tag{8.91}$$

Cooling of the medium results from the difference between the radiative fluxes out of and into it,

$$\dot{Q} = \dot{E}_L + \dot{E}_F - \dot{E}_P. \tag{8.92}$$

Defining the cooling COP to be $\kappa_{cool} \equiv \dot{Q}/\dot{E}_P$ and the optical-to-optical lasing efficiency as $\varepsilon_{lase} \equiv \dot{E}_L/\dot{E}_P$, one can then substitute (8.91) into (8.92) to obtain

$$\kappa_{cool} = \kappa_F(1 - \varepsilon_{lase}) - \kappa_L \varepsilon_{lase} \lambda_L/\lambda_F, \tag{8.93}$$

where the the cooling COP in the absence of lasing ($\varepsilon_{lase} = 0$) is $\kappa_F \equiv (\lambda_P - \lambda_F)/\lambda_F$, as in (8.9), and the lasing fractional quantum defect is $\kappa_L \equiv (h\nu_P - h\nu_L)/h\nu_P = (\lambda_L - \lambda_P)/\lambda_L$. The first expression on the right-hand side of (8.93) thus represents the relative cooling due to the fluorescence, while the second term is the relative heating due to the lasing. (Note in particular that $\kappa_{cool} = -\kappa_L$ if $\dot{E}_F = 0$.) For radiation balancing, $\kappa_{cool} = 0$ and (8.93) then implies that the optical efficiency is

$$\varepsilon_{lase} = \frac{\lambda_P - \lambda_F}{\lambda_L - \lambda_F}. \tag{8.94}$$

The choice of wavelengths consistent with the frequencies discussed above, $\lambda_L > \lambda_P > \lambda_F$, implies that $0 < \varepsilon_{lase} < 1$. Also note for this athermal case that the fraction of the pump power that is converted into fluorescence is $\dot{E}_F/\dot{E}_P = 1 - \varepsilon_{lase}$.

Even if perfect radiation balancing is not attained, one can reduce the heat load on the medium. For example, Bowman et al. [50] have constructed a quasi-cw thin-disk KGd(WO$_4$)$_2$ laser doped with 3 at.% Yb$^{3+}$ and edge-pumped with 90 diode laser bars (corresponding to an incident power of 2.25 kW) at $\lambda_P = 993$ nm. The highest output power $\dot{E}_L = 0.49$ kW at $\lambda_L = 1047$ nm resulted when $\dot{E}_P = 1.10$ kW of the pump power was absorbed, corresponding to an optical efficiency of $\varepsilon_{lase} = 45\%$. Using a powdered sample to minimize radiation trapping, the mean fluorescence wavelength was measured to be $\lambda_F = 997$ nm. (Therefore the pump wavelength, although a good match to an ytterbium absorption peak, was not long enough to produce athermal operation. It was limited by the available InGaAs laser diodes.) Consequently, one computes the nonlasing cooling COP to be $\kappa_F = -0.4\%$, and the lasing quantum defect to be $\kappa_L = +5.2\%$. Equation 8.93 thus predicts $-\kappa_{cool} = 2.7\%$, whereas the measured thermal loading was found to be 3.2%. The discrepancy results from the fact that the fluorescence wavelength is redshifted by reabsorption to $\lambda_F^* = \lambda_F/\eta_{ext}$ from (8.61), where the external fluorescence quantum efficiency is given by (8.81). Estimating the fluorescence lineshape function $g(\lambda)$ to be divided by $1 + \alpha_{res}(\lambda) L$ due to reabsorption, where $\alpha_{res}$ is the ytterbium absorption coefficient and $L$ is the effective length of the laser crystal (approximately double its 8-mm diameter) treated as an optical cavity, the trapped fluorescence wavelength is calculated to be $\lambda_F^* = 1011$ nm, which accounts for the extra thermal loading.

In principle, average athermal operation is possible not just for cw but also for pulsed lasers such as kilohertz Q-switched ytterbium-doped systems [51]. The idea is that net fluorescent cooling during the time that the Q-switch is off can compensate for a large transient thermal load when it is open, assuming that the pump source consists of continuous diode lasers. Another new concept [52] for mitigating the quantum-defect heating of a laser by optical cooling takes advantage of coherent anti-Stokes Raman scattering (CARS), in which a pump and a Stokes-shifted photon are converted into a pump and an anti-Stokes photon, with the associated annihilation of two phonons. As an example, a phase-matched cw silicon waveguide laser has been modeled that emits 0.69 W at a Stokes wavelength of 3.14 μm and 0.43 W at an anti-Stokes wavelength of 2.37 μm when pumped by a 5 W fiber laser at 2.7 μm. By increasing the ratio of anti-Stokes to Stokes photons, the simulations indicate that the thermal load is reduced by 35% compared to heating in the absence of CARS. An alternative to CARS in a single medium such as silicon is coherent four-wave mixing in a doped material such as $Yb^{3+}$:YAG [53]. Lasing occurs by Stokes shifting one pump photon at a dopant site, while cooling occurs by anti-Stokes shifting a second pump photon at a host site.

It will be interesting to see whether the first practical application of solid-state optical cooling outside of the laboratory will be to a refrigerator or to a laser. Time will tell.

## References

1 WINELAND, D.J. AND ITANO, W.M. (1987) Laser cooling, *Phys. Today*, **40**(6), 34–40.
2 PRINGSHEIM, P. (1929) Zwei Bemerkungen über den Unterschied von Lumineszenz- und Temperaturstrahlung, *Z. Physik*, **57**, 739–746.
3 VAVILOV, S. (1945) Some remarks on the Stokes law, *J. Phys.* (Moscow), **9**, 68–72.
4 PRINGSHEIM, P. (1946) Some remarks concerning the difference between luminescence and temperature radiation: Anti-Stokes fluorescence, *J. Phys.* (Moscow), **10**, 495–498.
5 VAVILOV, S. (1946) Photoluminescence and thermodynamics, *J. Phys.* (Moscow), **10**, 499–502.
6 LANDAU, L. (1946) On the thermodynamics of photoluminescence, *J. Phys.* (Moscow), **10**, 503–506.
7 ARAKENGY, A. (1957) Liouville's theorem and the intensity of beams, *Am. J. Phys.*, **25**, 519–525.
8 DJEU, N. AND WHITNEY, W.T. (1981) Laser cooling by spontaneous anti-Stokes scattering, *Phys. Rev. Lett.*, **46**, 236–239.
9 LIAKHOU, G., PAOLONI, S. AND BERTOLOTTI, M. (2004) Observations of laser cooling

by resonant energy transfer in $CO_2 - N_2$ mixtures, *J. Appl. Phys.*, **96**, 4219–4224.

10 MUNGAN, C.E., BUCHWALD, M.I. AND MILLS, G.L. (**2007**) All-solid-state optical coolers: History, status, and potential, in *Cryocoolers 14*, (eds S.D. Miller and R.G. Ross Jr.), ICC Press, Boulder, pp. 539–548.

11 KUSHIDA, T. AND GEUSIC, J.E. (**1968**) Optical refrigeration in Nd-doped yttrium aluminum garnet, *Phys. Rev. Lett.*, **21**, 1172–1175.

12 GEUSIC, J.E., SCHULZ-DUBOIS, E.O. AND SCOVIL, H.E.D. (**1967**) Quantum equivalent of the Carnot cycle, *Phys. Rev.* **156**, 343–351.

13 LANDSBERG, P.T., EVANS, D.A. (**1968**) Thermodynamic limits for some light-producing devices, *Phys. Rev.*, **166**, 242–246.

14 KAFRI, O. AND LEVINE, R.D. (**1974**) Thermodynamics of adiabatic laser processes: Optical heaters and refrigerators, *Opt. Commun.*, **12**, 118–122.

15 MUNGAN, C.E. (**2005**) Radiation thermodynamics with applications to lasing and fluorescent cooling, *Am. J. Phys.*, **73**, 315–322.

16 EPSTEIN, R.I., BUCHWALD, M.I., EDWARDS, B.C., GOSNELL, T.R. AND MUNGAN, C.E. (**1995**) Observation of laser-induced fluorescent cooling of a solid, *Nature* (London), **377**, 500–503.

17 Ross, R.T. (**1966**) Thermodynamic limitations on the conversion of radiant energy into work, *J. Chem. Phys.*, **45**, 1–7.

18 WEINSTEIN, M.A. (**1960**) Thermodynamic limitation on the conversion of heat into light, *J. Opt. Soc. Am.*, **50**, 597–602.

19 LANDSBERG, P.T. AND TONGE, G. (**1980**) Thermodynamic energy conversion efficiencies, *J. Appl. Phys.*, **51**, R1–R20.

20 LEHOVEC, K., ACCARDO, C.A. AND JAMGOCHIAN, E. (**1953**) Light emission produced by current injected into a green silicon-carbide crystal, *Phys. Rev.*, **89**, 20–25.

21 TAUC, J. (**1957**) The share of thermal energy taken from the surroundings in the electro-luminescent energy radiated from a p–n junction, *Czech. J. Phys.*, **7**, 275–276.

22 GERTHSEN, P. AND KAUER, E. (**1965**) The luminescence diode acting as a heat pump, *Phys. Lett.*, **17**, 255–256.

23 DOUSMANIS, G.C., MUELLER, C.W., NELSON, H. AND PETZINGER, K.G. (**1964**) Evidence of refrigerating action by means of photon emission in semiconductor diodes, *Phys. Rev.*, **133**, A316–A318.

24 NAKWASKI, W. (**1982**) Optical refrigeration in light-emitting diodes, *Electron Tech.* **13**, 61–76.

25 FOWLES, G.R. (**1989**) *Introduction to Modern Optics*, 2nd ed., Dover, New York, Chap. 7.

26 VAN BAAK, D.A. (**1995**) Just how bright is a laser?, *Phys. Teach.*, **33**, 497–499.

27 ESSEX, C., KENNEDY, D.C. AND BERRY, R.S. (**2003**) How hot is radiation?, *Am. J. Phys.*, **71**, 969–978.

28 NICODEMUS, F.E. (**1965**) Directional reflectance and emissivity of an opaque surface, *Appl. Opt.*, **4**, 767–773.

29 MUNGAN, C.E. (**2003**) Thermodynamics of radiation-balanced lasing, *J. Opt. Soc. Am. B*, **20**, 1075–1082.

30 CHEN, Y., RAPAPORT, A., CHUNG, T.-Y., CHEN, B. AND BASS, M. (**2003**) Fluorescence losses from Yb:YAG slab lasers, *Appl. Opt.*, **42**, 7157–7162.

31 RUAN, X.L., RAND, S.C. AND KAVIANY, M. (**2007**) Entropy and efficiency in laser cooling of solids, *Phys. Rev. B*, **75**, 214304:1–9.

32 CARTER, A.H. (**2001**) *Classical and Statistical Thermodynamics*, Prentice-Hall, Upper Saddle River, Chap. 20.

33 LOUDON, R. (**1990**) *The Quantum Theory of Light*, 2nd ed., Clarendon, Oxford, Chap. 1.

34 MUNGAN, C.E. AND GOSNELL, T.R. (**1999**) Laser cooling of solids. In: *Advances in Atomic, Molecular, and Optical Physics* (eds B. Bederson and H. Walther), Academic Press, San Diego, **40**, 161–228.

35 LEI, G., ANDERSON, J.E., BUCHWALD, M.I., EDWARDS, B.C., EPSTEIN, R.I., MURTAGH, M.T. AND SIGEL, G.H. JR. (**1998**) Spectroscopic evaluation of $Yb^{3+}$-doped glasses for optical refrigeration, *IEEE J. Quantum Electron.*, **QE-34**, 1839–1845.

36 THIEDE, J., DISTEL, J., GREENFIELD, S.R. AND EPSTEIN, R.I. (**2005**) Cooling to 208 K

by optical refrigeration, *Appl. Phys. Lett.*, **86**, 154107:1–3.

37 HOYT, C.W., HASSELBECK, M.P., SHEIK-BAHAE, M., EPSTEIN, R.I., GREENFIELD, S., THIEDE, J., DISTEL, J. AND VALENCIA, J. (**2003**) Advances in laser cooling of thulium-doped glass, *J. Opt. Soc. Am. B*, **20**, 1066–1074.

38 HOYT, C.W., PATTERSON, W., HASSELBECK, M.P., SHEIK-BAHAE, M., EPSTEIN, R.I., THIEDE, J. AND SELETSKIY, D. (**2003**) Laser cooling thulium-doped glass by 24 K from room temperature, *IEEE Conf. Proc. QELS Postconf. Digest*, 790–795.

39 BOWMAN, S.R. AND MUNGAN, C.E. (**2000**) New materials for optical cooling, *Appl. Phys. B*, **71**, 807–811.

40 WANG, J.-B., JOHNSON, S.R., DING, D., YU, S.-Q. AND ZHANG, Y.-H. (**2006**) Influence of photon recycling on semiconductor luminescence refrigeration, *J. Appl. Phys.*, **100**, 043502:1–5.

41 HEEG, B., DEBARBER, P.A. AND RUMBLES, G. (**2005**) Influence of fluorescence reabsorption and trapping on solid-state optical cooling, *Appl. Opt.*, **44**, 3117–3124.

42 LAMOUCHE, G., LAVALLARD, P., SURIS, R. AND GROUSSON, R. (**1998**) Low temperature laser cooling with a rare-earth doped glass, *J. Appl. Phys.*, **84**, 509–516.

43 SHEIK-BAHAE, M. AND EPSTEIN, R.I. (**2007**) Optical refrigeration, Nature Photonics **1**, 693–699.

44 EDWARDS, B.C., ANDERSON, J.E., EPSTEIN, R.I., MILLS, G.L. AND MORD, A.J. (**1999**) Demonstration of a solid-state optical cooler: An approach to cryogenic refrigeration, *J. Appl. Phys.*, **86**, 6489–6493.

45 CLARK, J.L., MILLER, P.F. AND RUMBLES, G. (**1998**) Red edge photophysics of ethanolic rhodamine 101 and the observation of laser cooling in the condensed phase, *J. Phys. Chem. A*, **102**, 4428–4437.

46 FREY, R., MICHERON, F. AND POCHOLLE, J.P. (**2000**) Comparison of Peltier and anti-Stokes optical coolings, *J. Appl. Phys.*, **87**, 4489–4498.

47 BOWMAN, S.R. (**1999**) Lasers without internal heat generation, *IEEE J. Quantum Electron.*, **QE-35**, 115–122.

48 PETRUSHKIN, S.V. AND SAMARTSEV, V.V. (**2003**) Laser cooling of active media in solid-state lasers, *Laser Phys.*, **13**, 1290–1296.

49 LI, C., LIU, Q., GONG, M., CHEN, G. AND YAN, P. (**2004**) Modeling of radiation-balanced continuous-wave laser oscillators, *J. Opt. Soc. Am. B*, **21**, 539–542.

50 BOWMAN, S.R., O'CONNOR, S.P. AND BISWAL, S. (**2005**) Ytterbium laser with reduced thermal loading, *IEEE J. Quantum Electron.*, **41**, 1510–1517.

51 WANG, X., CAO, D., ZHOU, M. AND TAN, J. (**2007**) Q-switched lasers with optical cooling, *J. Opt. Soc. Am. B*, **24**, 2213–2217.

52 VERMEULEN, N., DEBAES, C., MUYS, P. AND THIENPONT, H. (**2007**) Mitigating heat dissipation in Raman lasers using coherent anti-Stokes Raman scattering, *Phys. Rev. Lett.*, **99**, 093903:1–4.

53 MUYS, P. (**2008**) Stimulated radiative laser cooling, *Laser Phys.*, **18**, 430–433.

# Index

## a
absorption  4 ff
– bandgap engineering  171, 174
– bulk GaAs  155
– coefficients  4, 83
– cooling theory  151
– cryocoolers  119
– density of states engineering  172
– $Er^{3+}$-doped materials  103
– fluoride single crystals  76, 83
– luminescence  139 ff, 149
– metal-ion impurities  46
– microscopic theory  141 ff
– mirror heating  129
– rare-earth-doped materials  39
– thermodynamics  222
absorption–emission cycle  36
absorption–luminescence cycle  151
absorption–re-emission processes  93
acceptors  162
active ions  35–40, 48
actual performance  214 ff
aerosol dust particles  51
ammonium pyrrolidine dithiocarbamate (APDC) chelate  52 ff, 57 f
annihilation operators  141
antireflection coating  124
anti-Stokes fluorescence  1, 198
– design/applications  113, 117
– rare-earth-doped materials  33
– thermodynamics  228
anti-Stokes luminescence  75
applications  133–168
aqueous precursor route  49, 54 f
Auger recombination  140 ff
– bandgap engineering  178
– cooling theory  151
– GaAs quantum wells  159 ff
– surface plasmon polaritons  192
see also recombination

## b
background absorption  16, 21
– bandgap engineering  178
– cooling theory  151
– $Er^{3+}$-doped materials  112
– GaAs quantum wells  161
– metal-ion impurities  46
– thermodynamics  222
background dielectric constant  140
band blocking  13, 20
band diagrams  182
band-edge transitions  187
band-filling effects  157
bandgap energy  204
bandgap engineering  169–196
bandgap semiconductors see semiconductors
bandgaps
– $Er^{3+}$-doped materials  97
– shift/broadening  23
– structures  22
band-tail states  18, 175
see also Urbach tail
band-to-acceptor luminescence  163
basic concepts  1–33
$BaY_2F_8$ (BYF)  7, 35, 52
– Czochralski growth  66 ff
– fluoride single crystals  77–89
– thermal properties  41 ff
Beth–Uhlenbeck formula  146
BIG (fluoroindate) fabrication  65
binding energies  162, 172
black precipitates  45, 57
black-body radiation
– design/applications  119
– thermal load  11

– thermodynamics 208, 212
Bloch wave functions 176
blueshift 170, 193
Boltzmann statistics 9, 12, 18, 199 ff
Bose function 142
Bose–Einstein condensates 1
Bose–Einstein distribution 19, 205
break-even cooling condition 21
break-even nonradiative decay 16
break-even nonradiative lifetime 155, 161
break-even recombination 170, 178
Bridgman–Stockbarger crystal growth 65, 99
brightness temperature 203 ff, 212 ff
broadening 148, 174
bromide crystals 48, 68
bulk cooling
  – $Er^{3+}$-doped materials 105 ff
  – fluoride single crystals 89 ff
  – GaAs 153 ff
  – thermodynamics 197–225
bulk semiconductors 162 ff

## c

carbon impurities 162
Carnot coefficients 27, 197–225
casting 64 ff
cavities 11
cavity mirrors *see* mirrors
charge carriers 12, 139 ff
chelating agents 52 ff
chemical durability 40 ff
chemical potential 142–160, 176
chloride crystals
  – $KP_2Cl_5$ 77
  – net optical refrigeration 35
  – rare-earth-doped materials 68
chloride glass fabrication 48, 65 ff
chloride hydrolysis 98
chlorofluoride glasses 101
clean environment 51 ff
CNBZn (fluorochloride) glasses 65, 110
coefficient of performance (COP) *see* Carnot coefficients
coherent anti-Stokes Raman scattering (CARS) 230
collinear photothermal deflection spectroscopy 101
concentration quenching 40, 46
conductance 23

conduction band
  – GaAs quantum wells 159
  – luminescence 141 ff
  – type II quantum wells 182
conductivity 42, 77
contaminations 50, 78
cooling by anti-Stokes emission (CASE) 105 ff
cooling coefficients 200, 224
cooling efficiency 1 ff, 6, 14
  – bandgap engineering 169–196
  – bulk GaAs 156
  – design/applications 117
  – $Er^{3+}$-doped materials 101–115
  – fluoride single crystals 87 ff
  – luminescence 140
  – rare-earth-doped materials 36 ff, 40
  – rare-earth-doped solids 7 ff
  – surface plasmon polaritons 192
  – type II quantum wells 180
  – ZBLAN fabrication 63 ff
cooling elements 117, 131
cooling theory 151 ff, 162
copper impurities 10
correction factors 44
costs 136 ff
Coulomb interactions
  – bulk semiconductors 153, 162
  – GaAs quantum wells 159
  – luminescence 139 ff, 145
  – rare-earth-doped materials 37
Coulomb screening 13, 21, 152
creation operators 141
crucible materials 62
cryocoolers 1, 26
  – fluoride single crystals 75
  – modeling 119, 134 ff
crystal field splitting 10
crystal growth
  – $Er^{3+}$-doped materials 101
  – fluoride single crystals 77
  – rare-earth-doped halides 65 ff
  – ZBLAN 62
crystals 35
$Cs_3Tl_2C_{l9}$ crystals 66
Czochralski growth 65, 77 ff

## d

defects
  – bandgap engineering 188

– rare-earth-doped materials  36, 45
degeneracy factors  5, 38
degradation  117
density of states  171 ff, 204
density-matrix analysis  176
design
    – laser refrigerators  124 ff
    – rare-earth-doped materials  33–74
devitrification theory  59
dielectric constant  140 f
dielectric mirrors  117, 121
diethyl dithiocarbamate (DDTC) chelate  52
differential luminescence thermometry (DLT)  23, 35, 58
dipole matrix  142
direct gap semiconductors  12 f
dispersion  143, 190
dissolution  63 ff
distillation  55 ff
donor–acceptor transitions  171 ff, 183
doping  162, 172 ff
Doppler cooling  33, 197
drying  54 ff
Dyson equation  144

### e

effective temperature  200
efficiency see cooling –, quantum efficiency etc.
Einstein relations  87
electrochemical purification  57 ff
electroluminescent cooler  204
electromagnetic noise  134 ff
electron configurations  37
electron–hole pairs  13, 20
    – bandgap engineering  171
    – GaAs quantum wells  159 ff
    – luminescence  139–150
    – type II quantum wells  180
electron–hole plasma  145
electronic domains  169–196
electron–phonon interaction  5 f
emission  225
    – $Er^{3+}$-doped materials  109
    – fluoride single crystals  77
    – reabsorption  87
energy accumulation  222
energy bands  139

energy flux  206
energy gaps  75
energy level separation  4, 82
energy migration  39
energy transfer  189
energy transfer upconversion (ETU)  109 ff
engineering
    – bandgap  169–196
    – density of states  171 ff
entropy/–flux  197–212
epitaxial growth techniques  22
$Er^{3+}$-doped materials  97–116
erbium-doped fiber amplifiers (EDFAs)  34
escape probability  14
evaporative losses  63 ff
excited state absorption (ESA)  108 ff
excited states
    – $Er^{3+}$-doped materials  103, 109
    – hopping  119
    – population  221
    – rare-earth-doped materials  37
exciton density  146
exciton–exciton scattering  19
excitonic effects
    – bandgap engineering  176
    – cooling theory  151
    – GaAs  153, 161
    – quantum wells  161
excitonic resonances  141, 149
excitons–phonons coupling  148
expansion, thermal  43
experimentals
    – $Er^{3+}$-doped materials  101 ff
    – fluoride single crystals  78 ff
    – semiconductors  21 ff
external quantum efficiency (EQE)  15–27
external resonant cavity  11
extraction coefficient  160

### f

fabrication, rare-earth-doped materials  33–74
fast thermalization  110
Fermi fluid  153
Fermi functions  142 ff, 192
Fermi–Dirac distributions  12
fiberglass-epoxy support  119
filtration  51 ff
fining  63
fluorescence  7

- cryocoolers 117 f, 124 f
- design/applications 117
- fluoride single crystals 80, 87
- KPb$_2$Cl$_5$ crystal 98
- thermodynamics 198, 215
- type II quantum wells 183
- upconversion 1, 6

fluorescence quantum efficiency
- Er$^{3+}$-doped materials 103
- thermodynamics 215, 221, 226

*see also* quantum efficiency

fluoride crystals/glasses
- chemical durability 42
- net optical refrigeration 35
- rare-earth-doped materials 65 ff
- single crystals 75–96

fluorination 54 ff

fluorochloride glass
- CNBZn:Yb$^{3+}$ 34, 65
- Er$^{3+}$-doped materials 101 ff
- net optical refrigeration 35

fluoroindate glasses (BIG) 65

fluorozirconate glasses (ZBLAN)
- chemical durability 42
- purity 45
- Yb-doped 75–89

flux temperature 203, 212
four-level model 4 ff
free-carrier absorption (FCA) 17
full blackbody thermal load 11
full $T$ matrix theory 147
fundamentals 1–33
future rare-earth-doped materials 66 ff

## g

GaAs 17, 21 ff
- absorption spectra/photoluminescence 177
- cooling efficiency 17, 173
- luminescence 142
- photonic bandgap structure 187
- quantum wells 159 ff
- thermodynamics 204

GaN 189
gaps *see* band gaps
Gaussian profiles 206
glasses
- Er$^{3+}$-doped 101
- net optical refrigeration 35
- RE-doped 59 ff, 64 ff
- Yb$^{3+}$doped 7
- ZBLAN 64 ff
- ZrF$_4$ systems 59 ff

Green's function
- bulk semiconductors 156, 162
- luminescence 143–150

ground-state 9, 109
growth 22
- apparatus 78 ff
- Bridgman–Stockbarger 65 f, 99
- Czochralski 65 f, 77 ff
- epitaxial 22
- glass fabrication 65

## h

halide crystals/glasses
- Er$^{3+}$-doped 97
- rare-earth-doped 65 ff

Hamiltonian 140
hardness, fluoride single crystals 77
Hartree–Fock level 140, 144, 153
heat capacities 3 ff, 42
heat sink
- cryocooler 117 ff, 127 ff
- laser refrigerator 134
- microcoolers 136
- surface plasmon polaritons 189
- thermodynamics 197

heat transfer 119
heating by background absorption (HBA) 58
heat–light conversion 197
heavy metal fluoride glasses (HMFG) 34
- chemical durability 42
- fabrication 59

heavy metal halide systems 98
high-purity 21, 48 ff
*see also* purity
high-temperature limit 107
historicals 33 ff, 198 ff
holes density 163
homogenization 63 ff
host materials 7, 67
- cryocoolers 120
- Er$^{3+}$-doped 97 ff, 107
- fluoride single crystals 76, 92
- future trends 67
- rare-earth-doped 40 ff
- vibrational impurities 45

hot mirror 218
humid environments 42
hydrogen fluoride gas 54 ff
hydrolysis 42, 98
hydroxyl
    – fluoride single crystals 78
    – precursors 48, 54, 62
hygroscopic materials 42

*i*

impurities 10
    – bandgap engineering 174
    – fluoride single crystals 76
    – hydroxyl 48, 54, 62, 78
    – precursor fabrication 50
    – rare-earth-doped materials 36, 45 ff, 57 ff
indium-based layers 181
$InF_3$ oxidizers 61
infrared devices 35, 118
infrared transparency 97
interband absorption 13 f, 20
internal reflection 6, 11, 21, 26
    – mirror heating 130
    – rare-earth-doped materials 43
intra-laser-cavity 11
ion mobility 64
ionization 145, 172
iron impurities 10
irradiance 208
isotropic energy bands 141

*j*

Joule heating 204, 217

*k*

Keldysh contour 143, 164
kinetic holes 142
Knoop hardness 78
$KPb_2Cl_5$ crystal
    – Bridgman–Stockbarger growth 66
    – $Er^{3+}$-doped materials 98 ff
    – excitation spectra 110
Kramers–Kronig relation 143, 149
Kubo–Martin–Schwinger (KMS) relation 14, 142, 159
$KY(WO_4)_2:Yb^{3+}$ crystals 34

*l*

Lambert's law 206
laser cooling/refrigerating 1
    – design/applications 117–138

    – fluoride single crystals 75–96
    – rare-earth-doped materials 3, 33–74
    – semiconductors 12 ff, 134 ff, 169–196
laser dye solutions 139
laser excitation 4
lattice interactions 1, 6
lattice temperature 19
layers 162, 181
leakage 120 ff, 124 ff, 131
lifetime 134 ff
light extraction efficiency 117, 169
    see also extraction efficiency
light field, quantized 142
light propagation effects 150
light-emitting diodes (LEDs) 21
lineshape profile 208
Liouville's theorem, s 200
$LiYF_4$ (YLF) fluoride single crystals 77–89
load 2
longitudinal acoustic phonons 148
longitudinal optical phonons 148, 176
losses 19, 139
    – luminescence 150
    – ZBLAN fabrication 63ff
low phonon energy 97 ff
luminescence 21
    – bandgap engineering 174
    – $Er^{3+}$-doped materials 108
    – fluoride single crystals 87
    – microscopic theory 139–168
    – rare-earth-doped materials 37
    – thermodynamics 199
    – upconversion 1, 13

*m*

magnetic noise 134 ff
manybody interactions 12, 20
    – bandgap engineering 174
    – cooling theory 151
    – luminescence 139
mass, cryocoolers 134 ff
materials/–purity
    – $Er^{3+}$-doped 97–116
    – fluoride glasses 65 ff, 75–96
    – rare-earth-doped 33–74
matrix elements 171
maximum allowable nonradiative decay 16
McNamara–Mair devitrification 59

mechanical coolers 133
mechanical properties
  – fluoride single crystals 76
  – rare-earth-doped materials 40
melt quenching 61
metal halides 56
metal organic chemical vapor deposition (MOCVD) 22
metal-chelate complexes 52
metal–dielectric interface 189
metal-ion impurities 46 ff, 54
  see also impurities
methyl-isobutyl-ketone (MIBK) 52 ff, 57
microcooling applications 136 ff
microhardness 78
microscopic luminescence theory 139–168
mirrors
  – design/applications 117
  – dielectric 7, 10 f
  – heating 129 ff
  – hot 218
  – rare-earth-doped materials 43
  – thermodynamics 224
modeling/design 119 ff
modulation beams 103
molecular beam epitaxy (MBE) 25
momentum space 142
monochromatic light 1
monoclinic crystalline structures 77, 98
Mott transition 146, 149
multimode fiber-coupled laser 2
multiphonon decay 8
multiphonon relaxation
  – chemical durability 42
  – $Er^{3+}$-doped materials 97, 108
  – rare-earth-doped materials 37, 40 ff
  – thermodynamics 224
  – vibrational impurities 45
  see also relaxation
multiplets 37

## n

nanogaps 22, 26
narrowband light 197–213
net optical refrigeration 35
non-Lorentzian exciton lineshape 148
nonradiative recombination
  – bandgap engineering 169
  – cooling theory 151
  see also recombination
nonradiative relaxation
  – rare-earth-doped materials 33
  – thermodynamics 214
  – vibrational impurities 45
  – ZBLAN fabrication 62
  see also relaxation
nonradiative transitions 76
normalization 210
numerical aperture 130

## o

occupation number 18, 205, 211, 220, 224
open-batch processing 51
operating temperatures 42
optical cooling/refrigeration 202
  – cryocoolers 119
  – design/modeling 119 ff
  – microcoolers 136
  – net 35
  – solids 1 ff
  see also laser cooling
optical parametric oscillators (OPOs) 39
optical phonons 34
optical transitions 7 ff
optimum cooling rate 63 ff
oscillator strength 171
oxide crystals
  – chemical durability 42
  – fluoride single crystals 77
  – net optical refrigeration 35
  – rare-earth-doped 48
  see also YAG
oxide glasses 42

## p

parasitic absorption 3 ff, 10 ff, 151
  see also background absorption
parasitic heating 3 ff, 10 ff
  – fluoride single crystals 81
  – microcoolers 136
parasitic loads 120
partial entropy 211
Pauli exclusion principle 153
Peltier devices 8
performance coefficient see Carnot coefficients
periodically poled lithium niobate (PPLN) 39
phase-space filling 152

phenomenological cooling model 111 ff
phonons 1
– $Er^{3+}$-doped materials 97 ff
– fluoride single crystals 77
– GaAs quantum wells 159
– rare-earth-doped materials 34
– transitions 174 ff
photoluminescence 18, 178
photonic bandgap engineering 169–196
photons 2 ff
– cryocoolers 119
– cycles 139
– propagation 204
– tunneling 26
– waste recycling 27
photothermal deflection (PTD)
– cryocoolers 124
– $Er^{3+}$-doped materials 101 ff
– rare-earth-doped materials 35
photovoltaic converters 227
physical consistency conditions 149
physical properties 77 ff
Planck distribution 203 ff, 206, 211
Planckian thermal radiation 197
plasma fraction 145
plasma screening 153
polariton effects 151
polarization 85, 88, 124
population dynamics 103, 111
power density 12
power efficiency 123
ppb-level impurities 50, 57
precipitates 45, 60
precursors 48 ff
probabilities 120, 210
process equipment/conditions 50 ff
propagation constant 189
pulse tube coolers 133
pumping
– absorption 11
– bandgap engineering 179
– cryocooler modeling 128
– design/applications 117
– $Er^{3+}$-doped materials 102–114
– fluoride single crystals 87 ff
– luminescence frequency 139 f
– mirror heating 130
– resonant 108
– thermodynamics 202–227
– wavelengths 7
Purcell factors 189

purity/purification
– fluoride single crystals 78 ff
– $KPb_2Cl_5$ crystal 99
– precursor fabrication 50 ff
– rare-earth-doped materials 46–60

*q*
quantitative radiation thermodynamics 204 ff
quantized light field 142
quantum efficiency 4 ff
– bandgap engineering 179
– density of states engineering 171
– design/applications 117
– $Er^{3+}$-doped materials 103
– fluoride single crystals 76
– rare-earth-doped materials 37–45
– thermodynamics 198–226
– type II quantum wells 182
– vibrational impurities 46
quantum wells
– GaAS 159 ff
– type II 180 ff
quantum-confined systems 140
quasi-thermal equilibrium 142, 160
quenching centers 77

*r*
radiance 205
radiation temperatures 197
radiation thermodynamics 204 ff
radiative lifetime 182
radiative recombination 139 ff
rare-earth ions 37 ff
rare-earth-doped materials 12 ff
– $BaY_2F_8$ /$LiYF_4$ 7, 35, 41–68, 77–85
– design/fabrication 33–74
– glasses 139
– solids 3, 7 ff
reactive atmosphere processing 60
recombination 6, 13 f, 21
– cooling theory 151
– emission 204
– GaAs 153, 157 ff
– radiative 139 ff
recrystallization 51 ff
reddening effect 120
reflectance 125
reflection 6, 214

refractive index  19
– fluoride single crystals  77
– luminescence  142
– rare-earth-doped materials  40 ff
refrigeration technologies  133 ff
relaxation
– $Er^{3+}$-doped materials  97, 108
– rare-earth-doped materials  34 ff
– thermodynamics  214
– vibrational impurities  45
see also multiphonon relaxation
reliability  134 ff
renormalizations  145
retroreflectors  130
Roosbroeck–Shockley relation  14
ruggedness  134 ff

**s**
Saha equation  146
saturation  13
– bandgap engineering  173
– cryocooler modeling  122
– thermodynamics  223
scattering  45, 214
scheelite structure (YLF)  77
screened Hartree–Fock (SHF) model  154
second-order $T$ matrix  147
self-energies  140–156
SEM micrographs  26
semiconductors  2 ff
– bandgap engineering  169–196
– bulk  162 ff
– direct gap  12 ff
– efficiency  134 ff, 169–196
– luminescence  139–168
separation techniques  52
Shannon entropy  210
Sheik-Bahae–Epstein (SB–E) theory  153 f, 157 f
shielding  37
Shockley–Reed–Hall recombination  169
silver layers  124
simplified fluoride glasses  67 ff
single crystal fluorides  75–96
single-pass configuration  89 ff
$SnF_4$ oxidizer  61
sodium spectrum  198
sol–gel synthesis routes  49
solids  1–33

solvent extraction  52 ff
spacecraft applications  135
specific power  135
spectral energy flux density  206
spectroscopy
– $Er^{3+}$ upconversion properties  108 ff
– fluoride single crystals  80–86
spin–orbit interaction  37
spontaneous emission  3, 189
stability parameters  59
Stark sublevels  82
steady-state solution  5
Stefan–Boltzmann constant  210
Stirling's approximation  133, 210
sublimation  55 ff
sulfide glass fabrication  65 ff
surface plasmon polaritons  13, 189 ff
surface recombination  22
– bandgap engineering  169
– bulk semiconductors  162
see also recombination
susceptibility  145, 162
system mass  134 ff

**t**
temperature dependence  6
– Auger coefficients  152
– bandgap energy  148
– SB–E theory  155 ff
temperatures
– bulk GaAs  153
– bulk matter  197
– cryocooler modeling  120
– $Er^{3+}$-doped materials  106
– fluoride single crystals  76, 82, 89
– $T$ matrix  144 ff, 153
– ZBLAN fabrication  62
ternary phase diagram, ZBLAN  60
thermal energy accumulation  222
thermal link concept  2, 127
thermal properties  23, 77
– fluoride single crystals  76
– rare-earth-doped materials  40–45
thermalization  6, 20
– $Er^{3+}$-doped materials  110
– fluoride single crystals  87
thermocouples  35
thermodynamics  197–225
thermoelectric coolers (TECs)  8, 133

thermometry methods  35
thulium-doped materials  8
$Tm^{3+}$ fluoride crystals  67
$Tm^{3+}$ system (ZBLAN:$Tm^{3+}$)  34
total internal reflection (TIR)  42
trace impurity levels  57 ff
transition-metals  36, 46, 52
transitions  4, 7 ff
– bandgap engineering  171
– fluoride single crystals  76
– luminescence frequency  139
– type II quantum wells  182
translational invariance  159
transmission  214
transparency  77, 97
trapping  4 ff, 14
– bandgap engineering  169
– design/applications  117
– mirror heating  130
two-band luminescence model  141
type II quantum wells  180 ff

### u

ultratransparency  97
upconversion  108 ff, 139
Urbach tail  17
– bandgap engineering  174
– luminescence  148

### v

valence band
– bulk semiconductors  162
– GaAs  159
– luminescence  141
– type II quantum wells  182
van Roosbroeck–Shockley relation  14
vapor pressure  63
vertex corrections  147
vibrational impurities  45 ff
vibrational lattice modes  1, 107, 133 ff
viscosity optimum cooling rate  64 ff

### w

wavefunctions  37
wavelength-dependent temperature change  7, 91, 121

### y

$Y_2O_3$ oxide crystals  41
$Y_3Al_5O_{12}$ (YAG) crystals
– rare-earth-doped materials  33, 41
– thermal properties  43
$Yb^{3+}$ fluoride crystals  67
$Yb^{3+}$-doped fluorozirconate glass ($Yb^{3+}$:ZBLAN)  33
– design/applications  117
– optical cooler  208, 213
– thermodynamics  216, 220 ff
see also ZBLAN glass
$Yb^{3+}$ doping  7
$YLiF_4$  43
ytterbium-based optical refrigeration  1
ytterbium-doped fluorozirconate glass (ZBLANP)  7, 75

### z

ZBLAN glass
– chelating agents  53
– design/applications  117
– electrochemical purification  57
– fabrication  62 ff
– mirror heating  129
– precursor fabrication  48
– rare-earth-doped materials  7, 33–74
– ternary phase diagram  60
– thermal properties  43
– thermodynamics  208–224
see also $Yb^{3+}$:ZBLAN
zero-density exciton lineshape  148
zirconium  62
$ZnCl_2$-based glasses  65
ZnS/ZnSe  19, 22
zone refining  100
$ZrF_4$ glass formation  59 ff
$ZrOCl_2$ recrystallization  51